AF696501

PUBLICATIONS OF THE ISRAEL ACADEMY
OF SCIENCES AND HUMANITIES

CHAIM WEIZMANN

SCIENTIST, STATESMAN, AND
ARCHITECT OF SCIENCE POLICY

LEADERSHIP AND SCIENCE IN ISRAEL

CHAIM WEIZMANN

SCIENTIST, STATESMAN, AND ARCHITECT OF SCIENCE POLICY

Edited by

Benjamin Z. Kedar

Jerusalem 2017

The Israel Academy of Sciences and Humanities

English translation: Haim Watzman
Copy-editing and preparation for press: Deborah Greniman
Production: Yehuda Greenbaum

Photographs courtesy of
The Chaim Weizmann Archive, Yad Chaim Weizmann, Rehovot
The Weizmann Institute of Science Archive, Rehovot
The Hebrew University Central Archive

ISBN 978-965-208-225-1

©
The Israel Academy of Sciences and Humanities, 2017
Typeset by: Irit Nachum
Printed in Israel at Ayalon Printing, Jerusalem

Contents

Contents

Joshua Jortner

Preface

Dr. Chaim Weizmann, the first president of the State of Israel, was not only a statesman and a towering leader of the Jewish people, but also a notable scientist and an architect of academic policy and institutions for the state-in-the-making. His profound contribution to designing the contents and frameworks of scientific activity in the State of Israel and to their realization was unparalleled and has lasted until today.

In his inaugural address to the Knesset, Weizmann, the first President of the State of Israel, declared: "All my life, I have endeavored to make science and research a foundation for our national enterprise." These words faithfully reflect his belief that the State of Israel must be grounded, first and foremost, upon its spiritual, cultural, intellectual, and technological strength. This is our heritage: to see in science and scholarship a central dimension of our lives and a source of our strength, one that we must constantly nurture and develop, in the spirit of Weizmann's vision.

This volume is based on the lectures delivered at a conference held by the Israel Academy of Sciences and Humanities on January 8, 2013, to commemorate the sixtieth anniversary of Weizmann's passing. The Hebrew volume from which it was translated is the second in the Academy's series on "Leadership and Science in Israel." The first volume, published in 1989 and reprinted in 2013, was devoted to David Ben-Gurion.

We wish to thank the staff members of the various archives who helped us to locate and interpret the many photographs and documents that adorned the posters prepared for the conference and to produce the posters themselves: Ms. Merav Segal, Director of the Chaim Weizmann Archive; Ms. Keren Raz, then in charge of research relations and projects at the Chaim Weizmann Archive; Ms. Orna Zeltzer, then head of the Archive at the Weizmann Institute of Science; and Mr. Ofer Tzemach, Director of the Hebrew University Central Archive. Our thanks to Dr. Shaul Katz for preparing the archival materials displayed at the conference and to Ms. Avital Baer, the Academy's Spokesperson, for her assistance with mounting them for display and preparing them for printing. Our thanks as well to the staff of the Academy's Publications Department: to Ms. Tali Amir, its Director, who spared neither thought nor effort nor time in expertly preparing the Hebrew volume for press; to Ms. Deborah Grenimann, Senior Editor of English-Language Publications at the Academy, for editing the English version and preparing it for press; and to Mr. Yehuda Greenbaum, for supervising the production of both the Hebrew and the English volumes. Thanks are also due to Esther Rosenfeld, for proofreading the English volume, and to Nancy Zibman, for compiling the index. We are grateful, finally, to Mr. Haim Watzman for his fluent and erudite translation of the Hebrew articles into English.

Abbreviations

AEA	Albert Einstein Archives, The Hebrew University of Jerusalem
CWA	Chaim Weizmann Archive, Yad Chaim Weizmann, Rehovot
CWL	Meyer W. Weisgal et al. (eds.), *The Letters and Papers of Chaim Weizmann*, Series A: *Letters*, I–III, London–Rehovot: Oxford University Press–Yad Chaim Weizmann, 1968–1972; IV–VII, London–Jerusalem: Oxford University Press–Israel Universities Press, 1973–1975; VIII–XXIII, New Brunswick, NJ–Jerusalem: Transaction Books–Israel Universities Press, 1977–1980
CWP	Barnet Litvinoff (ed.), *The Letters and Papers of Chaim Weizmann*, Series B: *Papers*, I–II, New Brunswick, NJ–Jerusalem: Transaction Books–Israel Universities Press, 1983–1984
Reinharz, *CW*, I	Jehuda Reinharz, *Chaim Weizmann*, I: *The Making of a Zionist Leader*, New York–Oxford: Oxford University Press, 1985
Reinharz, *CW*, II	Jehuda Reinharz, *Chaim Weizmann*, II: *The Making of a Statesman*, New York–Oxford: Oxford University Press, 1993
Trial and Error	Chaim Weizmann, *Trial and Error*, London: Hamish Hamilton, 1949
WIA	The Weizmann Institute of Science Archive, Rehovot

Benjamin Z. Kedar

Introduction
Chaim Weizmann, Scientist and Statesman

While the phrase "scientist and statesman" is often applied to Chaim Weizmann, most of those who use it give no thought to how rare a combination it is. Very few scholars of Weizmann's time became political leaders, and the few who did came from the social sciences and the humanities, notable examples being U.S. President Woodrow Wilson, a political scientist, and Czech President Tomáš Garrigue Masaryk, a sociologist and philosopher. The most prominent case today is Germany's Chancellor Angela Merkel, a physicist, but she set aside her scientific work when she entered politics; similarly, Margaret Thatcher studied chemistry at Oxford and worked for a while as a research chemist, but she then decided to become a barrister and enter politics. Weizmann, who for many years worked simultaneously as a chemist and a Zionist statesman, was thus an exception. Perhaps only Benjamin Franklin combined science and politics in this way, but scientific research was pursued quite differently in Franklin's time than in Weizmann's. In sum, training in the natural sciences, which demands a direct and dispassionate expression of ideas, may not be the best preparation for the turbulence of the political world, where rhetoric plays such a central role. In Weizmann's case, however, science and politics may well have been able to cohabit precisely because he was a statesman without a state.

A reader of Weizmann's autobiography soon realizes that the tension between scientific work and Zionist activism is one of its central themes. Weizmann presented this tension as "the deep division of my life, or perhaps I should say its organic duality." He explained that the "tug-of-war between my scientific inclinations and my absorption in the Zionist movement has lasted throughout my life." The book chronicles the changing balance between those two forces over the years. During his studies in Berlin and Freiburg (Switzerland) he managed to keep them in equilibrium, but when he received his first academic appointment in Geneva, he devoted most of his time to Zionism, and, by his own admission, he was unable to focus on and fully carry out his scientific work.[1] His move to England in 1904 was in a sense an attempt to save himself from political over-involvement and to advance his scientific career. He largely succeeded, even if he did not manage to suppress his Zionist compulsion entirely.[2] The duality of his life was manifested also in the names he used during this period. As a Zionist, he was known as Chaim Weizmann, but he signed his scientific publications from 1904 to 1920, and all his patents, as Charles Weizmann.[3] During the early years of World War I he concentrated on his scientific work, with the aim of assisting the British war effort. Later in the war, however, "the centre of gravity of my life shifted once again towards my Zionist interests."[4] That is where they remained – from that time until his resignation from the presidency of the Zionist Organization in 1931, he set aside his scientific work almost entirely.[5] Afterward, especially after the establishment of the Sieff Institute in Rehovot in 1934, he found a way to combine his scientific and Zionist labors, putting

1 *Trial and Error*, pp. 76–77, 91. One of the drafts preserved in the Weizmann Archive shows that Weizmann first wrote "organic duality" and then replaced it with "conflicting dualism."

2 *Trial and Error*, pp. 123, 135, 137.

3 See the list of his scientific publications (and of his patents) in this volume. He also used the name Charles in two articles published in 1937 and 1938, while in one 1908 article he used the name Carl.

4 *Trial and Error*, p. 222. His wife Vera later recounted that when he returned from his visit to Palestine in October 1918, he told her: "Verotchka, I can't serve two mistresses – Science and Zionism." He maintained that in a year or so he would return to his laboratory. Vera Weizmann, *The Impossible Takes Longer*, New York–Evanston: Harper & Row, 1967, p. 89.

5 *Trial and Error*, p. 421.

his research in the service of building the Yishuv, the Jewish community in Palestine. But, time after time, events forced him to set aside his research.[6] During World War II, when he took part in the Allied war effort in Britain and the United States, he divided his time almost equally between science and Zionism,[7] thus restoring the balance of his early years. In the large picture, as Weizmann put it in a speech in 1945, "sometimes these two activities are not altogether compatible; not altogether, to use a chemical term, 'miscible.'" He did his best to apply his energies to both fields, but it was very difficult.[8]

Already during his studies in Berlin, Weizmann was aware that, as "only a student of chemistry,"[9] he was the exception among his Zionist friends, who were studying philosophy, history, economics, or law. He presumably also realized that he was not as good a speaker as he was a polemicist.[10] But Weizmann quickly learned to use his profession to leverage his Zionist activity. Kurt Blumenfeld, the leader of German Zionism, who first heard Weizmann speak at a public meeting in Berlin in 1906, testified that he spoke entirely differently from the other speakers. Weizmann declared:

> I am a chemist, and I therefore prefer to present Zionism and the Jewish question in the scientific terms with which I am familiar. Each one of you likes to eat a good meat soup; what we called in Yiddish *a Joich gemacht von der Mammen*. If it doesn't have enough

6 *Ibid.*, pp. 78, 422, 424–425.

7 *Ibid.*, pp. 520–521, 524.

8 Speech at a banquet of the American Committee for the Weizmann Institute of Science, New York, November 27, 1945. *CWP*, 2:83, p. 571.

9 *Trial and Error*, p. 54.

10 See: Dorothy Thompson, "The Mark of the Chosen," in Meyer W. Weisgal (ed.), *Chaim Weizmann: Statesman, Scientist, Builder of the Jewish Commonwealth*, New York: Dial Press, 1944, pp. 29–30; Israel Goldstein, "Leaves from a Diary," *ibid.*, p. 237; Leonard Stein, *The Balfour Declaration*, London: Vallentine Mitchell, 1961, p. 130; Abba Eban, "Tragedy and Triumph," in Meyer W. Weisgal & Joel Carmichael (eds.), *Chaim Weizmann: A Biography by Several Hands*, London: Weidenfeld & Nicolson, 1962, p. 250.

> salt, you add salt; if it is too salty, you throw it out. That, as I see it, is an analogy for the entire Jewish question in the countries of Europe. The non-Jews cook the soup, and we are the salt they need. They decide about the amount of salt, and woe to us if they taste us too strongly! Now the question arises: does the Jew want to be just the salt in someone else's soup, or does he want to constitute a substance of his own?

At the end of the meeting, Weizmann remarked smilingly to Blumenfeld (who would hear the same analogy many times in the future): "It is a good thing for a natural scientist to juxtapose his simple approach to the convoluted tangles of our philosophers."[11] Forty years later, during the tense political debate at the Twenty-Second Zionist Congress in Basel, Weizmann, speaking in Yiddish, tried to persuade his listeners that experimentation in the diplomatic sphere was worth pursuing. He cited his experience as a scientist:

> My friends! The work we are doing, the political work, is an empiric work, an experimental work. It is not a science; it is not an exact science. It is, as they say in English, a trial and error. You try something, sometimes one makes a mistake, sometimes it succeeds. I know what experimental works are and I know that you can do hundreds of experiments and they go badly and one goes well, and the one that goes well is tantamount to the hundreds that went badly. But when you do not do any experiments, nothing happens. That's why we must constantly keep trying, and to start again, and not get alarmed when something doesn't succeed.[12]

11 Kurt Blumenfeld, *Erlebte Judenfrage: Ein Vierteljahrhundert deutscher Zionismus*, Stuttgart: Deutsche Verlags-Anstalt, 1962, p. 40.

12 Speech by Dr. Ch. Weizmann, Minutes of the 22nd Zionist Congress, session 17, December 16, 1946, CZA LK 11/63, p. 3. A flowery Hebrew translation was printed in *The 22nd Zionist Congress, Basel, 9–24 December 1946: Stenographic Report*, Jerusalem: Zionist Organization Executive, n.d., pp. 340–341 (Hebrew). I am grateful to Chava Turniansky for her help in translating Weizmann's Yiddish.

It should be no surprise, then, that Weizmann chose to entitle his autobiography *Trial and Error*. Weizmann believed that the scientist in him helped him discern the state of the Zionist project clearly. He thus urged himself, just before his first visit to Palestine in 1907, to see the land "with the eyes of the chemist rather than those of the Zionist," even as he believed that historical and emotional attachment to the homeland would help Zionism overcome the obstacles observed by the chemist.[13]

That Weizmann was a scientist gave his pronouncements on Zionist affairs a special cachet in the eyes of others. In Jerusalem in March 1946, for example, one of the members of the Anglo-American Committee of Inquiry asked Weizmann if he, as a leading scientist, believed that it would be possible to bring 100,000 immigrants into the country within a single year. Weizmann offered a detailed response, weighing the different factors involved, and concluded that the immigration of that number of Jews was possible in the space of a year but perhaps might require two years.[14] Richard Crossman, one of the committee's members, was very much impressed by Weizmann's reply, which demonstrated, he thought, Weizmann's "curious scientific objectivity." The Zionist politicians present were agitated by Weizmann's ostensible concession, but his candid empirical approach impressed the committee's members far more than the routine Zionist propaganda speeches.[15]

What was the connection between Weizmann's scientific work and one of his greatest diplomatic achievements – the Balfour Declaration? In this regard, the account of David Lloyd George, the British prime minister at the time the declaration was issued, is well known: When he asked Weizmann what recognition he would like to receive for the

13 *Trial and Error*, p. 160.

14 *We Warned You, Gentlemen: Testimony of Dr. Chaim Weizmann before the Anglo-American Committee of Inquiry on Palestine*, Jerusalem: The Jewish Agency for Palestine, 1946, pp. 48–50.

15 Richard Crossman, *Palestine Mission: A Personal Record*, New York–London: Harper & Brothers, 1947, pp. 123–124. Four days after his testimony, Weizmann thought it best to send the committee a letter of clarification in which he emphasized that his intention had been that in any case the 100,000 Jews ought to be allowed to immigrate immediately.

scientific research that had been so valuable to Britain during the war, Weizmann replied that he wished nothing for himself; all he asked was to help the Jewish people return to the land of their ancestors. It was this request, Lloyd George wrote, that led to the Balfour Declaration.

Lloyd George first told this story in 1925, at an event held by the Jewish Historical Association in London,[16] at which, notably, Weizmann was present.[17] He repeated it in his memoirs, published in 1933, and at other opportunities, and many writers have taken his account at face value.[18] Yet it now transpires that the connection between Weizmann's work as a chemist and Britain's support of Zionism was made at least two months before the promulgation of the Balfour Declaration – and that the exchange upon which it turned was with Arthur James Balfour, not with Lloyd George. In a diary entry dated September 12, 1917, William Yale, a U.S. State Department emissary who stopped in London on his way to Cairo, mentions reports in British Jewish newspapers that a Jewish chemist had conveyed to the British government valuable secrets regarding the production of explosives. Upon meeting the chemist on a visit to the United States,[19] Balfour asked him what reward he desired. The chemist told him that he wanted nothing for himself, but

16 Philip Guedalla, *Napoleon and Palestine*, London: George Allen & Unwin, 1925, pp. 49–51.

17 Weizmann's remarks appear in *ibid.*, pp. 61–62. Weizmann did not relate to Lloyd George's story, but he responded to what the latter had said on Jewish–Arab relations in Palestine: "Mr. Lloyd George has touched upon the most important problem with which we are confronted. He will believe me when I say that the Jews are the last people in the world to begin to build their home on the back of somebody else. The Jews have suffered so much from injustice that they have learnt their lesson, and I can assure you that the Arabs will not suffer at our hands."

18 David Lloyd George, *War Memoirs*, II, London: Ivor Nicholson & Watson, 1933, pp. 584–587; Stein, *Balfour Declaration* (above, note 10), pp. 119–120; Jehuda Reinharz, *Chaim Weizmann,* II: *The Making of a Statesman*, New York–Oxford: Oxford University Press, 1993, p. 433, note 207.

19 Balfour's trip to the United States lasted from April 10 to June 9, 1917. See Kenneth Young, *Arthur James Balfour: The Happy Life of the Politician, Prime Minister, Statesman and Philosopher, 1848–1930*, London: G. Bell and Sons, 1963, pp. 380–384.

requested that, at the coming peace conference, the Allies give special consideration to the question of the Jews in Palestine. Balfour promised that this would happen. Yale added that the report had not been denied by the government, intimating that he believed it to be true.[20]

In this garbled 1917 version of the story, in which Weizmann and his meeting with Balfour are transposed to the United States, the essential element, the linkage between Weizmann's scientific work and British support for Zionism, is already there. Moreover, in 1922, three years before Lloyd George told his story in London, Ya'akov Ya'ari-Poleskin, a journalist who had spent a long time in the United States during the war, wrote that while there he had heard different versions of the "nice story about the Balfouric [*sic*] Declaration," according to which the English government had offered Weizmann many millions for his important inventions, "but Professor Weizmann turned down the millions and only asked that, in recompense, he be given the Land of Israel for the Hebrew nation."[21]

The factual foundation of the story is shaky. In April 1915, when Weizmann was first asked to work in the service of the British Admiralty, he indeed declared that he would forgo royalties until the war was

20 Yale recorded a "peculiar story recounted in Jewish papers of a chemist who gave certain secrets to the Government re. manufacture of explosives which was of great value, when Balflour [*sic*] was in America an interview was arrranged [*sic*] with this Jewish chemist who when asked what reward he desired replied that presonally [*sic*] he wished nothing but that he was a Jew and that he wished that in a Peace Conference the Allies would give special consideration to the question of the Jews in Palestine, which it was stated was promised, this article was never denied by the Government." Yale University Library, Manuscripts and Archives, William Yale Papers [M 658], Box 2, Folder 2 (Diary Notes, July–Sept. 1917). I am grateful to Michael Frost and Yale University for allowing me to examine a scan of this record. I was alerted to it by a partial quotation in Scott Anderson, *Lawrence in Arabia: War, Deceit, Imperial Folly and the Making of the Modern Middle East*, New York: Doubleday, 2013, p. 365. This book gives much space to Yale's varied activity during the war. I have not located the news report that Yale refers to in the British Jewish newspapers that I have been able to examine.

21 Ya'akov Ya'ari-Poleskin, *Dreamers and Fighters*, Petah Tikva: Sh.Z. Gisin, 1922, p. 183 (Hebrew). The author lived in the United States from 1915 to 1920.

over. But just a year later he was expecting payment for his work, and he was disappointed by the amount.[22] As for the conversation between him and Lloyd George, Vera Weizmann recounted many years later that Lloyd George's account was accurate.[23] But Weizmann was no doubt correct, in his autobiography, in rejecting Lloyd George's claim that the Balfour Declaration was personal compensation for his scientific services. "History does not deal in Aladdin's lamps," he asserted, referring to the magical lamp that, when touched, grants wishes.[24] Nevertheless, Weizmann's contribution to the British war effort undoubtedly aided him in gaining British support for Zionism's goals. It was thanks to his scientific work that he came into close contact with Lloyd George when the latter was serving as minister of munitions, as well as with other key figures, and that he became, to a certain extent, part of the British establishment. All this helped him in his contacts with these same people, leading to the Balfour Declaration.[25] During World War II, similarly, Weizmann's scientific work in the Allied cause helped him promote Zionist goals.[26]

Weizmann regarded laying the foundations for scientific research in the Jewish homeland as a manifestly political activity. At the Eleventh Zionist Congress, held in Vienna in September 1913, he declared: "There are moments in the lives of nations, and also in the life of the Jewish nation, in which [the establishment] of a school is also a

22 Jehuda Reinharz, "Science in the Service of Politics: The Case of Chaim Weizmann during the First World War," *English Historical Review*, 100 (1985), pp. 582, 597–599. See, in more detail, Reinharz, *Chaim Weizmann* (above, note 18), pp. 48, 61–67.

23 Stein, *Balfour Declaration* (above, note 10), p. 120, note 15.

24 *Trial and Error*, p. 192. On another occasion, a visibly annoyed Weizmann called Lloyd George's story "bilge" and forbade Weisgal to use it in Zionist work. See Weisgal, "Epilogue," in idem, *Chaim Weizmann: Statesman* (above, note 10), p. 339.

25 Stein, *Balfour Declaration* (above, note 10), p. 120; Reinharz, "Science in the Service of Politics" (above, note 22), pp. 601–603; Reinharz, *Chaim Weizmann* (above, note 18), pp. 68–72.

26 *Trial and Error*, pp. 519–520, 524, 527–531.

political act."[27] In January 1918, he asserted that there was an "urgent political reason" to purchase land on Mt. Scopus for the planned Hebrew University.[28] He made the founding of the university one of the goals of the Zionist Commission that was to set out for Palestine under the auspices of the British government in order to implement the Balfour Declaration.[29] In an article on the university published that year in Jerusalem, he explained (in Hebrew): "Under the conditions of the Exile, all cultural labor is also political and may well be the principal means of political work."[30] On July 24, 1918, in his speech at the laying of the university's foundation stones on Mt. Scopus – one of the most brilliant, if overlooked, speeches in the history of Zionism – he contrasted the destruction of the Temple and Jerusalem, marked just a few days previously on the ninth of the Jewish month of Av, with the new life bursting forth at the current ceremony, held on the fifteenth of Av, a day traditionally associated with rejoicing. He envisioned that the scientific work done at the university would rid the land of disease, restore the fertility of its soil, and demonstrate "the great value of the University in the building up of our national home."[31] Decades later he used the same language with regard to the Sieff Institute and the Weizmann Institute that grew out of it.[32]

How did Weizmann see the place of science in the building of the State of Israel, and what did he regard as the correct relationship between

27 *Stenographisches Protokoll der Verhandlungen des. XI. Zionisten-Kongresses in Wien, vom 2. bis 9. September 1913*, Berlin–Leipzig: Jüdischer Verlag, 1914, p. 158.

28 Weizmann to Jacobus H. Kann, January 20, 1918, *CWP*, 8, no. 76, p. 68.

29 See Benjamin Z. Kedar, "Laying the Foundation Stones of the Hebrew University of Jerusalem, 24th July 1918," in Shaul Katz and Michael Heyd (eds.), *The History of the Hebrew University. Origins and Beginnings*, Jerusalem: Magnes, 1997, p. 90, note 1 (Hebrew).

30 Chaim Weizmann, "On the Hebrew University," in *At the Threshold: A Collection on Matters of Life and Literature*, Jerusalem: Ahdut, 1918, pp. 16–17 (Hebrew).

31 See the photocopy of the original version of the speech in Kedar, "Laying the Foundation Stones" (above, note 29), pp. 110, 116–117.

32 *Trial and Error*, pp. 545–553.

science and the fundamental principles he upheld? In the address he delivered on February 14, 1949, at the opening of Israel's Constituent Assembly – which would soon constitute itself as the Knesset, the new country's parliament – he put it as follows, in a passage that deserves to be quoted in full:

> Let us endeavor first of all to expand our creative capacities by raising the standing of science and research in the State of Israel. Science and research are the foundations of all of human creation. We now must mobilize all the great things we have done in science and research in other countries for the building of our homeland. Yet, despite science's decisive importance, it alone will not save us. We must build a new bridge between science and the human spirit. "Where is no vision, the people perish" [Proverbs 29:18], and we have seen where scientific research not coupled with moral vision can lead – to the atom bomb that threatens to destroy the entire world and its inhabitants. All my life, I have endeavored to make science and research a foundation for our national enterprise. But I have also always known that above science there are sublime values and that only they can cure the ills of humanity – the values of justice and integrity, peace and brotherhood. "Zion shall be redeemed with justice and they who return to her with righteousness" [Isaiah 1:27].[33]

In his final years, Weizmann knew frustration and disappointment, in both the political and the scientific spheres. The "prisoner of Rehovot" did not enjoy his role as Israel's powerless president, forced to watch from the sidelines as the government pursued some policies that he considered misguided. He was also alarmed by the tendency, at the institute that would soon bear his name, to set aside pure science in favor of applied military work. That was probably the background to his call, at the ceremony inaugurating the Weizmann Institute of Science on November 2, 1949, to keep science "pure and untarnished."[34]

33 *Knesset Proceedings*, I, p. 7. Weizmann quoted this verse from Isaiah frequently.

34 Richard Crossman, "The Prisoner of Rehovoth," in Weisgal and Carmichael, *Chaim*

A few days before his death, he complained, in a conversation with Meyer Weisgal, about his helplessness in the face of the mistakes being made in Israel, and he voiced concerns about the future. Weizmann once more dwelt on the similarity between political and scientific conduct. "With their obstinacy," he observed, "[the Jews] will drive through a wall, but the breach in the wall (or as he put it in Yiddish, *der loch in der vant*) always remains gaping at you. Those who strive consciously to reach the mountain top remain chained to the bottom of the hill." Then he added: "Those who set out to achieve something specific in science, never achieve it. But those who work *lishmoh* (for its own sake) usually reach the top of the mountain."[35]

Weizmann (above, note 10), pp. 325–356; William B. Jensen, Henry Fenichel, & Milton Orchin, *Scientist in the Service of Israel: The Life and Times of Ernst David Bergmann (1903–1975)*, Jerusalem: Magnes, 2011, pp. 155–156, 166–169.

35 Meyer Weisgal, *So Far: An Autobiography*, New York: Random House, 1971, p. 278.

Joshua Jortner

Weizmann's Organic Chemistry in Academia and Industry

In 1962, David Ben-Gurion wrote: "Weizmann … was the man of intellect and of science, who not only understood the value of science in the resurgence of Israel and the creation of our new life, but who had himself made a tremendous contribution to scientific advancement, laying the foundations in this country for its institutions of higher learning – the Hebrew University in Jerusalem and the Weizmann Institute in Rehovoth."[1] It was an accurate expression of the central role that Zionism assigned to scientific research and humanistic excellence; both were considered vital elements in the reestablishment of the Jewish nation in its land. Weizmann asserted that the Jewish nation could attain its material needs only by developing its mental and intellectual capabilities.

In the history of Zionism and of the founding of the State of Israel, Chaim Weizmann stands among the greatest leaders that the Jewish people ever had. But Weizmann also stands out for his great scientific achievements in the fields of organic chemistry, biochemistry, and what we would now call biotechnology, which were of fundamental importance in the establishment of Israel's basic and applied research systems, and in determining the values and high level of Israeli science.

From the Zionist movement's infancy, scientists and intellectuals were its backbone. Among them were mathematicians Zvi Hermann

1 David Ben-Gurion, "Preface," in Meyer W. Weisgal and Joel Carmichael (eds.), *Chaim Weizmann: A Biography by Several Hands*, London: Weidenfeld and Nicolson, 1962, p. 5.

Shapira and Theodore Motzkin, botanist Otto Warburg, psychiatrist Max Nordau, and Martin Buber, a philosopher and scholar of Judaism. But Weizmann, a chemist, was the most central of these, and it was his vision and activity that formed the basis of the Israeli scientific community's values. He set that community's high standards and laid the foundation for its world-class areas of excellence. The relationship between Zionism and science was a central theme in Weizmann's Zionist consciousness. His scientific work was critical in shaping his identity and served his political and diplomatic work, which benefited from his prestige as the Jewish people's most outstanding scientist-statesman.

First Steps in the World of Chemistry

Writers of scientific biographies frequently seek the roots of their subjects' chosen scientific interest in the latter's adolescence. The young Chaim Weizmann attended a traditional Jewish religious school, a *ḥeder*, in Motal, the village of his birth. In 1886, at the age of eleven, he enrolled in a high school in the nearby city of Pinsk – on his own account, not a particularly good one, as its teachers were hired not on the basis of their pedagogical and intellectual abilities but rather thanks to their connections with the Russian bureaucracy. It was not an auspicious learning environment for Weizmann and the school's other talented students, most of whom were Jews.

But the school stood out in one field. Its chemistry teacher, a man named Kornienko, was a charismatic, intelligent, and gifted teacher with a broad background in science. According to Weizmann, it was Kornienko who opened up the world of chemistry for him. Years later, he wrote: "Very possibly, I owe [him] whatever I have been able to achieve in the way of science."[2] This may have been something of an exaggeration, because when Weizmann began his university studies in Germany he had to work very hard to catch up with his fellow students.

2 *Trial and Error*, p. 34.

Chemistry at the Turn of the Century

What was so exciting about the field of chemistry when Weizmann first encountered it in his high school in Pinsk?

The chemical sciences constitute the experimental and conceptual framework for exploring, on the molecular level, the structure, function, and changing of matter. The state of the field at that time is illustrated by the first book on chemistry to be published in Hebrew, in 1876, two years after Weizmann's birth. Its author was Zvi ben Meir Hacohen Rabinowitz, and it bore the lengthy and flowery title *Treasury of Wisdom and Science, including the Fundamentals of General Natural Wisdom and Also Information on Practical Manufacturing and Information on Various Sciences.*[3] The book surveys the preparation, atomic composition, and characteristics of materials, as well as the uses of these materials in the chemical industry. At the end of the nineteenth century, chemistry consisted almost exclusively of its organic and non-organic subfields. The modern study of the structure, activity, and transformation of materials – that is, of physical chemistry – began only during the first quarter of the twentieth century, while the foundation for theoretical chemistry was laid only in 1925 with the birth of quantum theory. The chemistry that Weizmann learned about from his high school teacher in Pinsk was thus confined to the synthesis and properties of organic and non-organic substances and their industrial applications. It was this world of experimental and applied science that Weizmann sought to join. His ambitions were motivated largely by the desire of a young Russian Jew to move up economically and socially.

Undergraduate in Berlin

After completing his high school studies with the highest marks (except in draftsmanship), Weizmann did not apply to Russian universities, which accepted only tiny numbers of Jewish students. Instead,

3 Vilna: Widow and Brothers Romm, 1876.

he turned to Germany, enrolling first at the Technical University of Darmstadt, where he spent a year, and then moving on to Berlin. In 1893 he matriculated in chemistry at Berlin's Royal Technical College of Charlottenburg, joining the laboratory led by Carl Theodore Liebermann (1842–1914), an organic chemist known for his work on dyes. Weizmann's specialization in organic chemistry was largely a matter of chance, determined by his professor's research. At the time, universities did not have chemistry departments that encompassed the field's full scope. Weizmann began to work under the direction of Karl Anton Augustin Bistrzycki (1862–1936), a Polish chemist who was a student of Liebermann's. Weizmann and Bistrzycki's work focused on the investigation of coal dust as a source of the organic dye alizarin, which constituted the basis of the molecular structure of dyes of considerable industrial importance. Weizmann worked with another student, Christian Deichler, on the preparation of synthetic dyes for textiles. Thus, at the beginning of his research career, in 1893–1897, Weizmann began to view chemistry as an important and exciting field of inquiry.

Graduate Work

In the autumn of 1897, upon completing his studies in Berlin, Weizmann moved on to the Catholic University of Freiburg in Switzerland, where his mentor, Bistrzycki, had been appointed to a professorship. It was only natural for Weizmann to follow his teacher and continue his scientific work toward his Ph.D. Bistrzycki shared his research area with Weizmann and suggested the topic on which Weizmann worked for his doctoral degree, under Bistrzycki's direction. Weizmann submitted his dissertation to the University Senate in 1899. It addressed two subjects:

1. The electrochemical reduction of anthraquinone. These were the early days of electrochemistry, an early stage of physical chemistry, developed by Michael Faraday in England and applied to the study of organic materials.

2. The condensation of phenanthraquinone and its derivative phenols. This part of Weizmann's work focused on classic organic chemistry, within his research adviser's larger project.

The title page of his dissertation identifies its author as "Chaim Weizmann aus Pinsk." Weizmann defended his work before a panel of examiners on January 17, 1899, and was awarded his doctorate *magna cum laude*.[4] Two memorial plaques are on display in the university's Senate chamber. One marks Weizmann's dissertation, while the other commemorates a Catholic priest who studied at the university and was murdered by the Nazis in the 1930s.

RESEARCH AND ZIONIST ACTIVISM AT THE UNIVERSITY OF GENEVA

From 1901 to 1904, Weizmann held the position of *Privatdozent* at the University of Geneva. His first research project to be published as an academic article, one that resulted in four patents, was carried out in cooperation with Christian Deichler, with whom he had first worked in Berlin. Reflecting Bistrzycki's influence, the project dealt with the synthesis of naphthacene quinone (a compound containing four aromatic carbon rings with two oxygen atoms) and its derivatives via the condensation of phthalic anhydride with naphthol by means of boric acid. The reaction, which came to be known as the Weizmann Reaction, was important in the chemical industry because it lowered the prices of

4 [In *Trial and Error*, p. 76, Weizmann writes that he graduated *summa cum laude*, with highest honors, but the official document from the University of Freiburg says *magna cum laude* (Mathematisch-Naturwissenschaftliche Fakultät der Universität Freiburg [Schweiz], Doktoratsregister 1899, pp. 14–15, No. 33). I am grateful to Zvi Rappoport of the Hebrew University of Jerusalem for providing me with a photocopy of this record, and to Mary Claire Berger of the University of Freiburg for providing the bibliographical reference for it. In the unpublished draft of his autobiography, Weizmann crossed out the passage regarding the grade given to his dissertation and replaced it with "I managed to pass my degree with high marks." See Illustrations 2 and 32 in this volume as well as Section V in the chapter "Unpublished Passages" – Ed.]

organic dyes by about thirty percent. It was Weizmann's first significant integration of pure and applied chemistry.

It was at this time that the tension between Weizmann's Zionist activity and his scientific work became evident. In March 1902 Weizmann gave his scientific work precedence over his Zionist labors, perhaps the only time he ever did so. Deep in his work and eager to display his achievements to the German chemical industry, he postponed a visit to London that was to have been devoted to putting in order the relations between the German and Geneva branches of the Zionist movement. That same year, however, Weizmann, Martin Buber, and Berthold Feiwel coauthored their famous document on the establishment of a Jewish university in Palestine.[5] This initiative, so important for the Zionist movement's science policy, came at the expense of Weizmann's scientific work. He faced the same conflict throughout his life.

Member of the Faculty at the University of Manchester

Weizmann's impressive scientific achievements – a prominent dissertation, five publications, and four patents – induced William Henry Perkin, who held the chair of organic chemistry at the University of Manchester, to offer the young chemist an academic appointment in his department. From 1904 to 1916, Weizmann served as a research assistant, demonstrator, and finally reader. In addition to his research and teaching responsibilities, he gave lectures in the service of science in his capacity as secretary of the Physical Society of Manchester, all while participating in Zionist affairs as well. Weizmann acquired many friends in his Manchester days who would later do groundbreaking scientific work. The most prominent of these was Robert Robinson (1886–1975), who would become one of the greatest organic chemists of the twentieth century, winning the Nobel Prize and serving as president of the British Royal Society. He had arrived in Manchester as a postdoctoral fellow, and he and Weizmann became close friends. Thanks to

5 Martin Buber, Berthold Feiwel, and Dr. Ch. Weizmann, *Eine jüdische Hochschule*, Berlin: Jüdischer Verlag, 1902.

this friendship, Robinson would become an active member of the Sieff Institute and later of the Weizmann Institute. Robinson's involvement exemplified a central tenet of Weizmann's scientific policy – using his personal and scientific connections to build a Jewish national research infrastructure.

Weizmann's scientific work at the University of Manchester is discussed elsewhere in this volume,[6] and I will therefore limit myself here to a few brief comments on his fields of scientific interest and activity. Weizmann was a gifted organic chemistry experimentalist who also evinced considerable interest in three other areas: theoretical chemistry (a field just then taking shape), biochemistry, and applied science. In 1907, Weizmann used empirical chemical theory to understand the colors of quinonic compounds by analyzing the absorption spectrum of organic materials. He collaborated in this pioneering work in the field now known as physical organic chemistry with the physical chemist E.C.C. Baly of Imperial College, London. On the applied side, in 1910 Weizmann made important contributions regarding the preparation and properties of synthetic rubber, working with the Curie Institute of Paris and the University of Manchester. In biotechnology, he worked on fermentation processes.

During his years of scientific work in Manchester, Weizmann became a leader in organic chemistry, focusing on polycyclic quinones (compounds containing multiple aromatic carbon rings that contain oxygen) and dyes composed of carbon compounds and other atoms. When he was appointed a reader in biological chemistry, Weizmann broadened his research from aromatic organic chemistry to aliphatic compounds. He now focused on materials of biological interest, such as amino acids and proteins. In 1911, he and Arthur Hopwood published an article on the synthesis of derivatives of amino acids, noting that "compounds of this type are of great importance in the field of physiology."[7] Another

6 [See Shaul Katz's chapter in this volume – Ed.]

7 Arthur Hopwood and Charles Weizmann, "Synthesis of Dipeptides of 7α-Aminolauric Acid with Glycine, Alanine, Valine, Leucine and Asparagine," *Journal of the Chemical Society*, 99 (1911), pp. 571–576.

innovative idea that he studied in 1911 was that amino acids could be the components of plant proteins. In 1913, he addressed the hydrolytic degradation of proteins and their separation into amino acids. Together with the well-known French chemist Victor Henri (1872–1940), he discovered the first photochemical degradation of an amino acid, glycine. These were all groundbreaking biochemical advances.

Scientific Research and Science Policy

Weizmann pursued important research on the photochemical degradation of amino acids in the 1930s. Simultaneously, at this dark time for the Jewish people, he worked assiduously to arrange refuge for Jewish scientists forced to flee Germany after the Nazi rise to power. Two of these were Nobel laureates in chemistry: Richard Willstätter (1872–1942), who discovered the structures of chlorophyll and hemoglobin, two molecules fundamental to life on earth, and Fritz Haber (1868–1934), who discovered a method of producing ammonia from hydrogen and atmospheric nitrogen that became fundamental to the manufacture of chemical fertilizers. Weizmann did the same for two scientists who would go on to receive Nobel Prizes in physics: Otto Stern (1888–1969), awarded the prize for his discovery and applications of molecular beams; and Eugene Wigner, awarded it for his work on the structure of the atomic nucleus. He also helped Ladislaus (László) Farkas (1904–1948), a physical chemist who worked on the internal dynamics of molecules; Fritz London (1900–1954), a physicist who discovered ultra-cold phenomena, such as superfluidity, and developed the first theory of chemical bonding; and Ernst David Bergmann (1903–1975), a promising organic chemist. Of these prominent scientists, only Farkas and Bergmann settled in Palestine, where they went on to found high-achieving chemical research frameworks: Farkas founded the outstanding physical chemistry department at the Hebrew University, while Bergmann laid the groundwork for prominent research in organic chemistry at the Sieff Institute and later at the Hebrew University. Bergmann was also instrumental in the establishment of Israel's military research and development program.

Weizmann's biochemical work laid the foundation for his critical discovery, in 1913, of a biochemical process for synthesizing acetone. This placed him in the vanguard of biotechnology and was the cornerstone of his standing as a scientist-statesman. He was also teaching at the University of Manchester at the time. The Weizmann Archives preserves a volume of his organic chemistry lectures to students from the 1908–1909 academic year through 1912. They are written in English, with marginal notes in German.[8] There are also examination papers from his organic chemistry and biochemistry classes (one typical question: "Describe the properties of a protein familiar to you from daily life").

The thirty-eight research articles that Weizmann published in professional journals during the early period of his scientific activity, up until 1915, are written in the best tradition of scientific writing – clear, concise, and objective. Another notable feature of these publications is the way Weizmann addressed errors that he discovered in the earlier work of other scientists. Many researchers tend, in such cases, to be blunt in their criticism. Weizmann always found a way to note politely that certain results reported by others required further examination. Louis Fieser, an organic chemist at Harvard, wrote: "Even in his formal scientific papers, some traces can be discerned of the warm human qualities and charm that have so contributed to his success in other fields of endeavor."[9]

8 See Illustration 6 in this volume.

9 Louis A. Fieser, "Investigations of Polycyclic Compounds and Physiologically Active Agents," in Meyer W. Weisgal (ed.), *Chaim Weizmann: Statesman, Scientist, Builder of the Jewish Commonwealth*, New York: Dial Press, 1944, p. 113.

Shaul Katz

Weizmann's Science Policy

I. Weizmann the Scientist: The Early Period

Chaim Weizmann took his first steps in science and scientific research in organic chemistry.[1] His introduction to the field in the academic-university context came in 1893–1897, at the Königliche Technische

Acknowledgments: I would like to thank those who helped me with good counsel and with assistance in locating archival material: Ms. Orna Zeltzer, Head of Archival Materials at the Weizmann Institute of Science; Ms. Merav Segal, Director of the Weizmann Archive at Yad Chaim Weizmann, and her colleague Ms. Keren Raz; and Mr. Ofer Tzemach, Director of the Central Archive of the Hebrew University of Jerusalem. I am also grateful to my wife Shifra Katz and my son Ariel Katz for helping me understand the sources cited here – the former with regard to the organic chemistry and microbiology documents, and the latter with regard to the patent documents. Thanks also to the English translator, Haim Watzman, and to my text and translations editor, Ruth Rossing.

1 This article relies heavily on materials in the Chaim Weizmann Archive (CWA), and on Weizmann's letters and papers, published, respectively, in *CWL* and *CWP*, including the introductions and notes provided by the series' editorial teams. The major events in Weizmann's life can be found in Appendix III (Concise Table of Dates) to the General Introduction by Leonard Stein, in idem (ed., in collaboration with Gedalia Yogev), *CWL*, I, pp. xxxvi–xxxvii.
The editorial board and editors chose not to include what they saw as "technical matters" concerning Weizmann's scientific work; see Stein, *ibid.*, pp. xx–xxi. Furthermore, the index does not include terminology relating to Weizmann's research. For example, in his letter of December 15, 1935, to Richard Willstätter (*CWL*, XVII, no. 102, pp. 93–94), Weizmann reports on his new experiments on butyl alcohol fermentation, but neither "butyl alcohol" nor "fermentation" appears in the volume's index. The editors later changed their approach and decided that material relating to Weizmann's scientific work would be published in a special Third

Hochschule zu Berlin.[2] Under the direction of his teachers there, he specialized in synthetic organic chemistry and its industrial applications.

Series; see Meyer Weisgal, "Introduction," *CWL*, VI, p. 21. However, Ernst David Bergmann, who was to have been in charge of the Third Series, had little time to devote to the project, and it was left on the doorstep of Yad Chaim Weizmann. The information concerning Weizmann's academic employment in 1904–1916 at the University of Manchester has been augmented by a document referred by University of Manchester Academic Secretary Vincent Knowles on January 9, 1967, to Gedalia Yogev; see CWA, Scientific Division, box 26, Manchester file. The document lists Weizmann's academic appointments, his attendant responsibilities, and his salary. See also Weizmann's letter of July 1, 1906, to his fiancée, Vera Khatzman, in which he informs her that he has received a Master's degree from Manchester and will receive his doctorate – D.Sc. – shortly before Christmas: *CWL*, IV, no. 279. I am grateful to Gabriel Shapira for examining the original text on my behalf. The D.Sc. degree was awarded to Weizmann (under the name Charles Weizmann) in 1907. I am grateful to Dr. Eran Tal of the Chaim Weizmann Archive for this information, which he obtained from James Peters, an archivist at the University of Manchester Archive, in a letter dated August 29, 2014. In 1926, the response of the academic secretary of the newly founded Hebrew University to an official British Mandate questionnaire provides information on the university's principal officers and scholars. Regarding Weizmann, it states: "Chaim Weizmann Sc.D. (Freiburg) D.Sc. (Manchester)." See S. Ginzberg, Registrar to Universities Bureau of the British Empire, November 18, 1926, Hebrew University Archive, 1, Overseas Universities, 1925–1930.

2 Today the Technische Universität Berlin. A brief survey of this institution's history and transformations can be found in the introduction by its president, Eberhard Knobloch, to *"The Shoulders on Which We Stand": Wegbereiter der Wissenschaft – 125 Jahre Technische Universität Berlin*, Berlin–Heidelberg: Springer, 2004, pp. v–xii. Technological universities were a new phenomenon in the academic world of the second half of the nineteenth century. For a concise account of this development, see Christophe Charle, "Patterns," in Walter Rüegg (ed.), *Universities in the Nineteenth and Early Twentieth Centuries (1800–1945)*, III: *A History of the University in Europe*, Cambridge, UK: Cambridge University Press, 2004, pp. 57–58. Institutions of this sort should not be confused with "polytechnics," as Stein mistakenly did in his General Introduction to *CWL*, I (above, note 1). On the polytechnic as a post-secondary educational institution devoted largely to training students for industry, see the glossary in Asa S. Knowles et al. (eds.), *The International Encyclopedia of Higher Education*, I, San Francisco–Washington–London: Jossey-Bass, 1977, p. 484a.

His principal advisor was Karl Anton Augustin Bistrzycki (1862–1936), a student of Carl Theodore Liebermann (1842–1914), who had been a student of Johann Friedrich Wilhelm Adolf von Baeyer (1835–1917, Nobel laureate 1905). Von Baeyer was one of the founders of synthetic organic chemistry and among the first to "artificially" or "synthetically" produce natural substances in the laboratory.[3] One of the most important of these was indigo, the synthesis of which, in 1878, was a landmark achievement in the field, as was the production of synthetic alizarin from anthracene by Carl Graebe (1841–1927, see below) and Liebermann that same year.[4] In addition to his contribution to academic research, von Baeyer and his associates and students were involved in the founding of the modern organic chemical industry.[5]

3 See Bistrzycki's obituary, including a list of his scientific works, in Constantin Gyr, "Augustin Bistrzycki (1862–1936)," *Helvetica Chimica Acta*, 20/1, pp. 477–489, http://onlinelibrary.wiley.com/doi/10.1002/hlca.19370200174/pdf (in German). Carl Theodor Liebermann was a highly regarded teacher and researcher at his institution; see *Die Technische Hochschule zu Berlin, 1799 –1924: Festschrift*, Berlin: Georg Stikle, 1925, pp. 92, 94. Nevertheless, the editors of this anthology of celebrated scientists at the Technische Hochschule did not see fit to include Liebermann. He headed the chemistry and metallurgy department and taught courses in organic chemistry; see *Königliche Technische Hochschule zu Berlin, Program für das Studienjahr 1895–96*, Berlin 1895, pp. 6, 66. Bistrzycki, a *Privatdozent*, was responsible for practical laboratory training in organic chemistry; *ibid.*, p. 68. Liebermann's genealogy reveals family ties with the painter Max Liebermann and German Foreign Minister Walter Rathenau (assassinated in 1922); see www.deutsche-biographie.de/sfz51145.html. I am grateful to Dana Bruheller for directing my attention to this and similar sources. An entry devoted to Carl Liebermann can also be found in *Jüdisches Lexikon*, III, http://sammlungen.ub.uni-frankfurt.de/freimann/content/pageview/364760 (all accessed January 29, 2017).

4 J[ames] R[iddick] Partington, *A History of Chemistry*, IV, London: Macmillan, 1964, pp. 755–788 (von Baeyer), 788–790 (Graebe), 790–793 (Liebermann). On the chemical genealogy of von Baeyer and his students – the "Baeyer tree" – see http://careerchem.com/NAMED/TREES-IDEAS/Baeyer1.pdf (accessed January 29, 2017).

5 For more on von Baeyer and Liebermann in this context, see Carsten Reinhardt & Anthony S. Travis, *Heinrich Caro and the Creation of Modern Chemical Industry*, Dordrecht: Kluwer Academic Publishers, 2000, pp. 179–186.

Such association of commercial and academic research activity was considered acceptable academic practice in the field of synthetic organic chemistry. In other areas of research at the time, however, such as mathematics at the University of Berlin, researchers embraced an ideal of pure research, not only rejecting any inclination to endeavors outside the academic world but also repudiating those engaged in them. Nevertheless, many members of the scientific elite of this period, at least in Berlin, viewed what we would today see as the tension between basic research and applied research as a cornucopia for industrial development.[6] Weizmann, as will be discussed below, adopted this position and endeavored to institutionalize it.

At the Königliche Technische Hochschule Weizmann received the title "Technischer Chemiker,"[7] but he presumably sought an academic career. To that end he followed his teacher, Bistrzycki, when the latter was invited to assume a professorship at the University of Freiburg (Switzerland). There, in 1899, Weizmann submitted his dissertation and

6 This tradition of "pure mathematics" was brought to the Hebrew University of Jerusalem with its founding and was fostered there for many years. On its origins in Berlin and those who brought it to Jerusalem, see Shaul Katz, "Berlin Roots – Zionist Incarnation: The Ethos of Pure Mathematics and the Beginnings of the Einstein Institute of Mathematics at the Hebrew University of Jerusalem," *Science in Context*, 17/1–2 (March–June 2004), pp. 199–234. Nevertheless, the tradition of pure mathematical research was not widespread. The Physikalisch-Technische Reichsanstalt in Charlottenburg, built adjacent to the Königliche Technische Hochschule, made a novel organizational connection between academic science and research, directed to the needs of industrial manufacturing. It was founded in 1887 by the celebrated physicist Hermann von Helmholtz and the no less celebrated industrialist and entrepreneur Werner von Siemens. This institution functioned as what we would today call a government physics laboratory and national standards institute, providing professional services to industry. Among its administrators were theoretical physicists, including Albert Einstein and Max Planck, while experimental physicists, including Emil Warburg, worked in its laboratories. See David Cahan, *An Institute for an Empire: The Physikalisch-Technische Reichsanstalt, 1871–1918*, Cambridge, UK: Cambridge University Press, 1989, pp. 59–125, 180–185, 200–212; a map showing the close proximity of the two institutions, as well as the nearby University of Berlin, appears on p. 90.

7 Knowles to Yogev (above, note 1).

was awarded his doctorate *magna cum laude*.[8] Only then, after he had joined Graebe, did Weizmann move on to the University of Geneva, where he served as Graebe's assistant from 1899 to 1904[9] and was also appointed a *Privatdozent* – an independent teacher whose earnings depended on the number of his students – considered the first step up the academic appointments ladder in the German-speaking world at that time.[10] In 1904, when he realized that he would apparently not receive any further academic promotions in Geneva, he followed Graebe's recommendation that he go to England.[11] He was accepted into the department led by William Henry Perkin, Jr. (1860–1929), at what was then called Victoria University of Manchester.[12] The year 1904 thus ushered

8 See Illustration 2 in this volume. At the University of Manchester the degree from Freiburg was referred to as "Sc.D."; see Knowles to Yogev (above, note 1). Nevertheless, years later Weizmann wrote: "I still find myself under the necessity of filling out the lacunae in my education which should have been attended to forty and fifty years ago. And I look with envy on young colleagues whose scientific education is much sounder than mine" (*Trial and Error*, p. 91). On the topic of his dissertation, see Illustration 1 and the first entry in the list of his scientific publications in this volume. The topic is also noted in Gyr's list of Bistrzycki's publications in 1899 (above, note 3), p. 485.

9 See the sources cited in note 2 above. Graebe was also a student of von Baeyer, and, as noted above, his success in synthesizing alizarin is considered on a par with von Baeyer's synthesis of indigo. Weizmann's decision to join Graebe may thus be seen as evidence of the beginning of Weizmann's ascent up the academic ladder in the field of synthetic organic chemistry of the time. As late as 1911, when Weizmann sought a highly respected scientist to recommend him, he suggested Graebe. See Weizmann to Vera Weizmann, *CWL*, VI, no. 236 (April 8, 1911).

10 On the status of a *Privatdozent*, see Joseph Ben-David, "The Profession of Science and its Powers," in *Scientific Growth: Essays on the Social Organization and Ethos of Science by Joseph Ben-David*, ed. Gad Freudenthal, Berkeley: University of California Press, 1991, pp. 196–197.

11 See Jehuda Reinharz, *CW*, I.

12 Perkin was also a student of von Baeyer and had worked alongside him in the years 1882–1886. See Jack Morrell, "W.H. Perkin, Jr., at Manchester and Oxford," *Osiris*, 8 (1993), pp. 104–126. The first encounter between Perkin and Weizmann was described by Robert Robinson, who was present and later remarked on Perkin's satisfaction with the new recruit. See idem, *Memoirs of a Minor Prophet:*

in a twelve-year period in which Weizmann was associated with the University of Manchester; joining what would today be termed a center of scientific excellence, he conducted research, taught, rose in the academic ranks, and concomitantly worked for the chemical industry. But despite his doctorate and his rank of *Privatdozent* in Switzerland, during his initial years in Manchester he had to earn academic qualifications all over again and obtain a master's degree and doctorate in science from the University of Manchester itself.[13]

Manchester 1904–1916

1. The Synthetic Organic Chemistry Period

From the end of the nineteenth century, the University of Manchester, ostensibly a "provincial" British institution, enjoyed rapid expansion and was home to some of the greatest scientists of the early twentieth century. These included physicists, chemists, physical chemists, and organic chemists.[14] Among the physicists were such renowned figures as Arthur Schuster (1851–1934) and Ernest Rutherford (1871–1937,

70 Years of Organic Chemistry, I, Amsterdam: Elsevier, 1976, p. 46. Eventually, Weizmann recruited Robinson to the Weizmann Institute's Board of Trustees and as a member of the Institute.

13 Weizmann was first made a research student; in January 1905 he became a research fellow, and he later served in other positions and was promoted to laboratory demonstrator. The University of Manchester granted Weizmann its own degrees, after appropriate procedures – first the M.Sc. and then the D.Sc.; see Weizmann's 1906 letter to Vera Khatzman (above, note 1). On the 1909 faculty roster of the chemistry department, Weizmann appears as "Charles Weizmann, D.Sc." with the title "Senior Lecturer and Assistant Director of the Organic Laboratories." See "Staff of the Chemical Department," CWA, Scientific Division, box 26, Manchester file.

14 Robert H. Kargon, *Science in Victorian Manchester: Enterprise and Expertise*, Manchester: Manchester University Press–Johns Hopkins University Press, 1977, esp. pp. 153–257; Mary Jo Nye, *From Chemical Philosophy to Theoretical Chemistry: Dynamics of Matter and Dynamics of Disciplines: 1800–1950*, Berkeley: University of California Press, 1993, esp. pp. 177–182 (Weizmann and his laboratory are mentioned in passing on p. 180).

Nobel laureate 1908), who helped bring down the barriers between chemistry and physics both theoretically and on the academic teaching level.[15] Working in Perkin's laboratory were chemists who defined a new type of organic chemistry, such as Arthur Lapworth (1872–1941)[16] and Robert Robinson (1886–1975, Nobel laureate 1947). Lapworth was one of the founders of physical organic chemistry and later succeeded Perkin as head of the department – a post that Weizmann, who saw himself as a natural candidate, was extremely disappointed not to receive.[17] Robinson began his career in synthetic organic chemistry with the complex synthesis of alkaloids, but over time, like Lapworth, he turned his attention to the reaction mechanisms of organic molecules, developing an electronic theory of organic reactions, and other aspects of theoretical organic chemistry.[18] But the originality evident in their

15 On Schuster see G.C. Simpson, "Sir Arthur Schuster, 1851–1934," *Biographical Memoirs of Fellows of the Royal Society,* I, no. 4, p. 408, DOI:10.1098/rsbm.1935.0006 (http://rsbm.royalsocietypublishing.org/content/royobits/1/4/408, accessed November 28, 2011); and *Trial and Error*, pp. 148, 151–152. On Rutherford see Kargon, *Science in Victorian Manchester*, pp. 214–237; and Nye, *From Chemical Philosophy to Theoretical Chemistry*, pp. 163–195 (both above, note 14).

16 Lapworth held chairs in both physical chemistry and organic chemistry at the University of Manchester; see Robert Robinson, "Arthur Lapworth, 1872–1941," *Obituary Notices of Fellows of The Royal Society,* V, no. 15 (February 1947), pp. 555–572 (www.jstor.org/stable/769101, accessed November 28, 2011).

17 Reinharz writes about this from Weizmann's point of view, as well as about his disappointment at not being elected to the Royal Society, despite his expectation that he would be; see Reinharz, *CW*, I, pp. 361–367. Vera Weizmann, who presumably voiced her husband's position, wrote: "The professorship he had hoped to succeed to on Professor Perkin's retirement went to Professor Lapworth, Perkin's brother-in-law, an *inorganic* chemist" (emphasis in the original). See Vera Weizmann, *The Impossible Takes Longer: The Memoirs of Vera Weizmann, Wife of Israel's First President, As Told to David Tutaev*, New York–Evanston: Harper & Row, 1967, p. 39. But family relationships certainly do not explain why Weizmann was not elected to the Royal Society. Lapworth was elected in 1910.

18 Robinson's scientific biography, like those of other Nobel Prize laureates, can be found on the Nobel Prize website (www.nobelprize.org/nobel_prizes/chemistry/laureates/1947/, accessed November 28, 2011); for his autobiography, see above, note 12.

research programs, like those of others working at Manchester, was not reflected in Weizmann's research programs; his principal innovation lay in bringing new types of feedstock into the synthetic organic chemistry arena. Instead of, or alongside, the usual cracking and decomposition products of coal, coal tar, and oil, he introduced the products of biological fermentation processes derived from farm crops. No less important, Weizmann succeeded in transferring such processes from the laboratory to industry.

At Manchester, Weizmann found himself in direct contact with scientific greatness, much more so than he had experienced in Berlin, Freiburg, or Geneva. Moreover, at Manchester Weizmann was welcomed into the social circle of the leading local scientists, who made up, or would soon make up, part of the world's scientific elite. Rutherford, for example, invited Weizmann to his monthly luncheons, attended by the cream of Manchester's scientific and scholarly elite.[19] This must certainly have broadened Weizmann's views on the nature of scientific research of all kinds. He was exposed to "the spirit of the new century," namely, the belief in optimism derived from the symbiosis between science and modern society.[20] Weizmann later made use of these early connections with the elite of British science for the research institutions he founded in Palestine.[21] Furthermore, as his biographers have noted,

19 Among them the young Niels Bohr; see John Campbell, *Rutherford: Scientist Supreme*, Christchurch, NZ: AAS Publications, 1999, p. 342. See also David Wilson, *Rutherford: Simple Genius*, London: Hadder and Stoug, 1983, pp. 225–227, 482. Weizmann made no mention of these gatherings in his memoirs. By contrast, he reported to his wife about meeting some members of the French scientific elite a few years after he gained entry into the exclusive circle of Manchester scientists: "I am simply delighted by the French School of Physics and Chemistry. They are very brilliant people, and extremely interesting. How can one compare them with our ponderous mediocrities, or even the Germans." Weizmann to Vera Weizmann, *CWL*, VI, no. 293 (April 4, 1914), p. 314.

20 Kargon was referring to an editorial published in *Nature* at the dawn of the twentieth century: *Nature*, 63/1627 (January 1901). See Kargon, *Science in Victorian Manchester* (above, note 14), pp. 236–237.

21 For example, Arthur Schuster, who "left part of his splendid library to the Hebrew University"; see *Trial and Error*, p. 152.

it was during his time in Manchester that Weizmann established himself as a leading British Zionist.

Having begun at Manchester in the junior position of assistant to Perkin, Weizmann gradually assumed greater responsibilities at the university, became more independent in the teaching tasks assigned him, rose in the academic ranks, and became an admired teacher.[22] After immersing himself in bacteriology, he began to integrate that field into organic chemistry teaching – a true innovation in the definition of the traditions and practices in those fields. In 1913 he was given a laboratory of his own, signaling his total academic independence.[23] That same year he was promoted to the rank of reader in biochemistry.[24]

Over the years, Weizmann was able to refine the research program he brought with him to Manchester, and his publications were highly regarded and attracted much interest, as attested by references to them made by scientists at the time as well as years later.[25]

22 Horace B. Speakman, "Dr. Weizmann's Contribution to Microbiology," in Paul Goodman (ed.), *Chaim Weizmann: A Tribute on His Seventieth Birthday*, London: Victor Gollancz, 1945, pp. 130–136. Speakman describes Weizmann's lectures with admiration. See also Weizmann's own reports on the esteem in which his students held him, in his 1906 letter to Vera Khatzman (above, note 1). On Weizmann's progress at the University of Manchester see above, note 13.

23 In 1913, when Perkin left the university, Weizmann took over Perkin's laboratories and received a private office and storeroom for his equipment. The biochemical laboratories were given the status of a sub-department, under Weizmann's direction. See Jehuda Reinharz, "Science in the Service of Politics: The Case of Chaim Weizmann during the First World War," *English Historical Review*, 100 (1985), p. 576.

24 At the time, the academic ranks at the University of Manchester were: Assistant Lecturer, Lecturer, Senior Lecturer, Reader, Professor. See Knowles to Yogev (above, note 1).

25 For a contemporary reference see, e.g., Edward de Barry Barnett, *Anthracene and Anthraquinone*, New York: Van Nostrand, 1921. Louis Fieser wrote in 1944: "Weizmann's first chemical research resulted in a major discovery of such a scope that fruitful applications and elaborations of the basic principle of the discovery have been made during the ensuing forty years, by Weizmann and others, without exhausting the possibilities." Idem, "Dr. Weizmann's Research in the Polycyclic Series," in Goodman, *Chaim Weizmann: A Tribute* (above, note 22), p. 113.

Weizmann was also able to cultivate new research programs not directly connected to those of the Liebermann-Perkin school. It seems reasonable to suppose that it was his chemistry instruction of medical students that led him to turn to what could at that time have been considered biochemistry.[26] Along these lines, he began, in 1911, to investigate amino acids and proteins, transitioning from research that primarily involved the synthesis of dyes from feedstock to the study of materials of biological importance. But Weizmann's sortie into biochemistry was short-lived; it produced but a single article and did not constitute a new research direction, at least for him.[27]

At Manchester, Weizmann encountered others who shared his "chemical genealogy." As he wrote in his autobiography, "It so happened that the work I had been doing under Graebe had been on similar lines to that which the father of Professor Perkin, Sir William Henry Perkin, had done nearly half a century before."[28] Can Weizmann's reflection on his scientific genealogy be taken as testimony that the tradition of synthetic organic chemistry research in which he was working had passed its peak of revolutionary innovation?[29]

26 Speakman on Weizmann's laboratory: "Work on the constitution of proteins was in progress; the classical methods of Emil Fischer were used." Idem, "Dr. Weizmann's Contribution" (above, note 22), p. 131. Fieser also addressed the connection between Weizmann's projects and biochemistry: see idem, "Dr. Weizmann's Research" (above, note 25).

27 The article in question was Charles Weizmann & Ganesh Sakharam Agashe, "Hydrolysis of Proteins with an Alcoholic Solution of Hydrogen Chloride, Part I," *Biochemical Journal,* 7/5 (October 1913), pp. 437–440. The volume number indicates just how new the field was.

28 *Trial and Error*, p. 126. In other words, if Weizmann's teachers Liebermann and Graebe belonged to the second generation of the tradition begun by von Baeyer and William Henry Perkin, Sr. (1838–1907), the first to produce synthetic aniline, then Weizmann, as a synthetic organic chemist, belonged to the third or even the fourth generation.

29 The literature on scientific change, such as Thomas Kuhn's concept of paradigm shifts and Imre Lakatos's work on changes in research programs and their conceptual systems, seems inappropriate to the case at hand, which would appear to concern the exhaustion of the innovative academic potential of a particular

2. On to Bacteriology!

Weizmann's principal innovation at Manchester was to depart from the Liebermann-Perkin tradition. He was the first to demonstrate the possibility of integrating bacteriological knowledge with synthetic organic chemistry.[30] Actually, at first Weizmann had no intention of creating a new interdisciplinary research agenda. He had no comprehensive research program – that would come only later. He would write in his autobiography that he was drawn to biological chemistry and bacteriology a few years after his arrival in Manchester, as a specialty within organic chemistry. He dated his first interest in the idea to a few years after his arrival in Manchester,[31] though his vision of a program of biotechnological research would emerge only two decades later, at the beginning of the 1930s, in the period leading up to the establishment of the Daniel Sieff Research Institute.[32]

How did Weizmann become a bacteriologist? He began to visit the Pasteur Institute in Paris no later than March 1909, where he trained himself in bacteriological research methods, with an emphasis on fermentation processes.[33] This turn toward bacteriology led to his most im-

research program. In any case, to conceptualize Weizmann's professional story and its transformations would take us far from the subject under discussion here.

30 "Zymotechnic knowledge" was the term used at that time for what we today refer to as biotechnology. See Robert Bud, *The Uses of Life: A History of Biotechnology*, Cambridge, UK: Cambridge University Press, 1994. Weizmann did not use the term "zymotechnic," and he seems not to have seen himself as belonging to that tradition. This was not the case with Speakman, his student and partner in fermentation research at Manchester, who transferred the study of fermentation to the acetone factories of Toronto (see below). From the beginning of the 1920s, and during his tenure at the University of Toronto, Speakman's academic appointment was as professor of zymotechnology. I am grateful to Harold Averil, assistant archivist at the University of Toronto, for giving me access to archival material on Speakman.

31 *Trial and Error*, p. 171.

32 See Chapters II and III of this article.

33 Weizmann to Vera Weizmann, *CWL*, V, nos. 188–192 (March 23, 25, 27, 29, and 31, 1910). Weizmann also became acquainted with medical bacteriology; see Weizmann to Vera Weizmann, *ibid.*, no. 193 (April 5, 1910); and *Trial and Error*, pp. 170–171.

portant discovery. As Louis Fieser wrote in 1944, "Chaim Weizmann's discovery of an efficient microbiological method for the large-scale production of acetone and butanol … stands out as the most striking accomplishment in the chemical career of this gifted and versatile man."[34] Later, with the revival of interest in biotechnology and its derivatives, such as biological fuel, Weizmann would be recognized as one of the field's pioneers.[35]

Weizmann left us no real account of how he became interested in bacteriology and its possible affinity to synthetic organic chemistry. In his autobiography, he writes: "It was during this period, too, that I began the study of fermentations. I was led to this subject by its relation to the production of synthetic rubber."[36] But Weizmann's involvement in synthetic rubber began in 1910, while his active entry into bacteriology began in the winter of 1909 at the Pasteur Institute, making it difficult to determine which came first, or whether the two were connected in his thoughts, perhaps even a few years before. Some sixty years later, Vera Weizmann – quoting her husband's succinct remark, "Let nature and bacteria work for me and produce synthetic rubber!" – dated this interest to the time of her arrival in Manchester, in 1906.[37] Organic chemists and historians of the field have raised a number of conjectures as to the origin of the idea, ranging from the influence of other scientists he

34 Fieser, "Dr. Weizmann's Research" (above, note 25), p. 113.

35 See D.T. Jones & D.R. Woods, "Acetone-Butanol Fermentation Revisited," *Microbiological Reviews*, 50 (December 1986), pp. 484–524 (www.ncbi.nlm.nih.gov/pmc/articles/PMC373084/, accessed November 28, 2011); and Pazhamalai Anbarasan et al., "Integration of Chemical Catalysis with Extractive Fermentation to Produce Fuels," *Nature*, 491 (November 8, 2012), pp. 235–239. [See also the article by Raphael Lamed and Edward A. Bayer in this volume. – Ed.]

36 *Trial and Error*, p. 171. With the exception of the patent he received in 1915 for fermentation and the production of acetone (see below), at this time Weizmann had not yet written any public scientific document offering an indication of his attitude toward his predecessors. Weizmann's written testimony closest in time to the 1915 patent is "Dr. Weizmann Technical Statement Dated 2nd February 1925 (supplemental to brochure)," CWA, 15-983. See also note 41 below.

37 See Vera Weizmann, *Impossible Takes Longer* (above, note 17), p. 34.

knew personally, to the influence of professional literature,[38] to totally fortuitous circumstances relating to Weizmann as a Zionist interested in promoting economic development in Palestine.[39] As against these, or alongside them, we cannot rule out the possibility that Weizmann was stimulated to think in terms of bacteriology and fermentation as a result

38 Ernst David Bergmann maintained that Weizmann was influenced by work on yeast fermentation carried out by Manchester biochemist Arthur Harden. See E.D. Bergmann, "Obituary Notice: Chaim Weizmann, 1874–1952," *Journal of the Chemical Society* (1953), pp. 2840–2844.

39 Robert Bud, "The Zymotechnic Roots of Biotechnology," *British Journal for the History of Science*, 5/1 (March 1992), pp. 137–138 (www.jstor.org/stable/4027008?seq=2, accessed November 29, 2011). From Reinharz, *CW*, I (p. 302), Bud learned of documents relating to a short exploratory trip Weizmann made to Palestine in the summer of 1907 at the behest of the Jewish industrialist Johann Kremenetzky, for the purpose of examining the feasibility and advisability of establishing industries based on the country's agricultural produce. Bud links this trip to Weizmann's turn toward microbiology in relation to fermentation. While the letters Weizmann sent to Kremenetzky indeed address manufacturing based on the industrial use of agricultural products, they include no direct reference to the production of industrial goods by means of fermentation in other than traditional processes, if at all. See, e.g., Weizmann to Kremenetzky, *CWL*, V, nos. 28 and 60 (April 19 and September 26, 1907). Weizmann's suggestion of "taking the production of olive oil from its point of origin and manufacturing soap" (as the Atid factory, founded in Haifa in 1919, did) is the clearest evidence that Weizmann not only did not propose any biotechnological ideas to Kremenetzky, but in fact chose, as an example of the kind of manufacturing that Palestine needed, a project based on modern technology. See Weizmann to Kremenetzky, *ibid.*, no. 64 (November 17, 1907). Further evidence of the lack of biotechnological ideas in the plan Weizmann submitted to Kremenetzky was the list of ideas for manufacturing materials via fermentation that Weizmann drew up six years later in a letter to his friend Julius Simon, a prominent British Zionist. Without any connection to the future of industry in Palestine, Weizmann proposed the production of lactic acid, butyric acid, fusel oil, and other kinds of alcohols by "fermentation techniques"; see Weizmann to Julius Simon, *CWL*, VI, no. 83 (June 14, 1913). On the technological history of the Atid (later Shemen) factory, see Shaul Katz, "On the Technological History of Palestine in the Late Ottoman Period: Three Case Studies," in Israel Bartal (ed.), *The Second Aliya: Studies*, Jerusalem: Yad Izhak Ben-Zvi, 1998, pp. 189–212 (Hebrew).

of a chance conversation – perhaps, for example, with his acquaintances Moise and Rachel Schoen, family friends of Rachel (Vera's sister) and Joseph Blumenfeld, who lived in Paris. Moise Schoen worked at the Pasteur Institute as an assistant and eventually a partner of Auguste Fernbach (1860–1939).[40] In a relatively late reference, Weizmann cited bacteriological research literature as a source of his interest in the field.[41]

40 This is the opinion of John Polanyi, an organic chemist born in 1929 who received the Nobel Prize in 1986. He gave a lecture in 1998 entitled "The Instructive Tale of 'Weizmann's Bacterium,'" based no doubt on an oral tradition that reached him when he was a student in Manchester. "Through this [his marriage to Vera]," Polanyi said, "he acquired a scientist brother-in-law in Paris, with a professional interest in fermentation. His imagination stimulated, Mr. Weizmann, a synthetic organic chemist, decided to find out how good bacteria might be at synthetic organic chemistry. He soon found that they were a lot more skillful than him. They were able to make something called butyl alcohol out of sugar." See http://sites.utoronto.ca/jpolanyi/public_affairs/public_affairs4j.html (accessed November 29, 2016). Polanyi references Lionel Davidson, *The Sun Chemist*, London: Jonathan Cape, 1976, an imaginary biography reflecting what the author heard during time he spent in Rehovot. See the unnumbered note by the author on the last page of that book.

41 The circumstances described below regarding the development of the acetone production process led Weizmann to register a patent, and a patent applicant naturally stresses his or her sole proprietary hold on the process in question. Weizmann did not, however, supplement his patent registration with an article written in the scientific research genre in which the claim to an innovative research discovery comes with a survey of "precursor" research that led up to the reported findings. Apparently, the circumstances of the scaling up of the laboratory process into an industrial technology, along with Weizmann's decision to leave the university and devote himself more intensely to political life, left him with no time to write a professional article intended for scientific journals. A report on such previous work and the sources of bacteriological knowledge from which Weizmann took his ideas can be found in his "Technical Statement" (above, note 36). References to the work of his predecessors and to appropriate research literature can be found, for example, in his mentions there of Pasteur's 1861 discovery of butanol as a waste product of food production and of Fitz's research in the years 1879–1882 (cf. Jones & Woods, above, note 35), and in his citation of "*Les Anaérobies* by [M.] Jungano & [A.] Distaso, Paris, 1910, p. 25." But these testimonies by Weizmann about himself are problematic. The document in question is part of a memorandum for the

Whatever the case, by 1910 Weizmann was already familiar with bacteriological fermentation processes in the laboratory and with Fernbach's expertise in this field. It was Weizmann's initiative to enlist Fernbach and his prowess in Perkin's synthetic rubber project.

3. From the Synthetic Rubber Project to the Acetone Project

At the beginning of the twentieth century, the growing demand for automobiles, and thus for tires, led to a shortage of caoutchouc – natural rubber – sparking initiatives for the industrial production of artificial rubber. It was known that natural rubber consists of large molecules – polymers – made up of chains of smaller monomer isoprene molecules, each containing five carbon atoms. To produce synthetic rubber, it was necessary to solve two problems. First, isoprene or an analogous material had to be produced in its monomer form, to serve as the raw material for industrial production of the polymer. Second, a way of stringing the monomers together into a polymer had to be found. Strange and Graham, Ltd., an English company, was one firm that took up the challenge of producing synthetic rubber.[42] In 1910, the company invited Perkin to join the new project and help find a solution to the first problem, the production of isoprene or a similar monomer that could be synthesized into a polymer chain with properties like those of natural rubber.

In the tradition of synthetic organic chemistry, Perkin and his research group first proposed to find a synthetic process for producing the natural compound that was in short supply. In other words, they sought to find the most efficient chemical process for producing isoprene monomers.[43] But, thanks to his knowledge of the bacteriological

defense in a lawsuit filed against Weizmann and others in the wake of the success of his acetone project (in the end, the judge recognized the "Weizmann bacterium" as Weizmann's intellectual property). Reinharz discusses this suit at length. See Reinharz, *CW*, II, pp. 40–72 and passim.

42 Bud, in discussing Weizmann and the Weizmann Process, also addresses the rubber crisis and the individuals and companies that sought to solve it. See Bud, *Uses of Life* (above, note 30), pp. 37–48.

43 See the testimony of Harold Davies, who worked on the project: idem, "A Review

research tradition, Weizmann proposed that Fernbach be brought from Paris into the project, and that the industrial production of synthetic rubber be based on materials taken from bacteriological fermentation. Weizmann's recommendation was the beginning of a process that led him to the discovery of an efficient microbiological method for the large-scale production of acetone and butanol.[44]

What was the innovation in Weizmann's integration of the two fields? Traditionally, fermentation caused by bacteria and other microorganisms had been used in the production of foodstuffs – such as bread, beer, wine, and cheese – from agricultural products which, in their raw form, are either not fit for human consumption or are of better quality or more easily preserved in their fermented form. These processes create byproducts that were long seen as useless or even as waste. This was the status of butanol and acetone when they were first discovered, in the context of bacteriological research. Louis Pasteur is credited with isolating butanol as a byproduct of bacteriological fermentation in 1861, while acetone, along with acetic acid and ethanol, were identified as bacteriological fermentation byproducts by Franz Schardinger in 1904 – discoveries now considered two of the most important events in the history of microbiological science. As far as Pasteur and Schardinger were concerned, however, butanol and acetone were worthless; they did not divine these substances' potential for agriculture or industry. It was Weizmann who realized their value. His Patent 4,845, which he registered in Britain in 1915 under the title "Improvements in the Bacterial Fermentation of Carbohydrates and in Bacterial Cultures for the Same,"[45] centered on the identification of the bacterium that generates fermentation to break down starch into acetone and alcohols. Like Fieser, the members of the American Society for Microbiology, in their

of the Development of Butyl Alcohol-Acetone Fermentation and Early Work on Synthetic Rubber," in Israel M. Sieff (ed.), *Papers Collected to Commemorate the 70th Anniversary of Dr. Chaim Weizmann*, [n.p.], 1944, pp. 5–12. Davies would later accompany Weizmann to London and then to Rehovot. A brief biography of Davies can be found in the Biographical Index in *CWL*, V.

44 Fieser, "Dr. Weizmann's Research" (above, note 25), p. 113.

45 [See no. 23 in the list of Weizmann's patents in this volume. – Ed.]

list of "Significant Events in Microbiology 1861–1999," assign a high value to Weizmann's discovery, placing it alongside those of Pasteur and Schardinger.[46]

How did this change in Weizmann's research agenda come about? His working conditions at the University of Manchester enabled him to pursue both research and consulting outside the university,[47] and his consulting work allowed him to register and sell patents.[48] Although Weizmann would later come under criticism for having devoted too much time to such work outside the university, in fact the institution viewed its scientists' efforts on behalf of industry as perfectly legitimate, as was the case at the other universities where Weizmann had been employed previously. The patents he registered jointly with Perkin are proof of this.[49] The cooperation between the two must certainly have been the background to Perkin's invitation to Weizmann to join the ambitious project for the production of artificial rubber.

Perkin took up Weizmann's idea that butanol, a four-carbon molecule that was a byproduct of bacteriological fermentation, could be used to produce butadiene, an analogue of isoprene, which could be used to produce synthetic rubber. Weizmann then suggested bringing Fernbach and his assistant and partner Schoen into the project. Fernbach, an expert in bacterial fermentation, was director of the Pasteur Institute's Service des fermentations, and it was in that capacity that Weizmann probably became acquainted with his work.[50] Weizmann's

46 www.asm.org/index.php/choma3/71-membership/archives/7852 (accessed November 29, 2016), entry for the year 1915.

47 Knowles to Yogev (above, note 1).

48 For example, "Preparation of *iso*-Bornyl Esters from Pinene Hydrochloride," no. 21 on the list of Weizmann's patents appended by Fieser to his article "Dr. Weizmann's Research" (above, note 25), is a 1909 German patent held jointly by Weizmann and Clayton Aniline Co.

49 [See nos. 18–22 in the list of Weizmann's patents in this volume. – Ed.]

50 On Fernbach, see Albert Delaunay, *L'Institut Pasteur: Des origines à aujourd'hui*, Paris: Editions France Empire, 1962. Fernbach is mentioned there a dozen times, and Schoen half as many. On pp. 113–114, 185, and 298 they are mentioned together. See also "Auguste Fernbach, 1860–1939," *Annales de l'Institut Pasteur*,

intention must certainly have been to use the fermentation of biological raw materials to obtain the building blocks needed for the production of artificial rubber.

What Weizmann did not anticipate, however, was that bringing Fernbach into the project would be the first link in a chain of events that would lead him, a few years later, to the acetone project. In 1911, during his first year on the project, Fernbach was able to isolate a bacterium that ingested potato starch and excreted butanol. This was an important achievement, but Perkin's group was not particularly pleased. Potatoes were a problematic raw material because of the cost of transporting them and because of their storage requirements. Furthermore, the process's yield was insufficient. The team wanted a microorganism with a broad profile that could produce butanol from a wide range of vegetable sources containing starch.

Dissatisfaction with the limitations of Fernbach's discovery was accompanied by increasing mutual suspicion, intrigues, and disagreements among the researchers, including sharp differences between Weizmann and Perkin. They vied over the division of intellectual property and future profits from the products of their research, and both were in conflict with the Strange and Graham company and the factories that had been recruited to develop manufacturing processes based on the scientific research. In the end the partnership broke up, and Weizmann left the synthetic rubber project.[51]

However, Weizmann continued to investigate bacterial fermentation processes on his own.[52] He soon isolated a new species of bacterium, or

62/3 (March 1939), pp. 249–252, which contains a list of Fernbach's publications; beginning in 1910, Schoen was a co-author of about half of them. Fernbach's employment in the synthetic rubber project and Weizmann's work in Fernbach's laboratory in 1912 comprise about a third of the brief biography of Fernbach included in a collection of such biographies on the website of the Pasteur Institute: "Auguste Fernbach (1860–1939)," www.pasteur.fr/infosci/archives/f-bio.html (accessed November 29, 2016).

51 Reinharz addresses this affair and its protagonists at length. See idem, *CW*, II, pp. 40–72.

52 Weizmann's friend Julius Simon helped fund his research. On the goals of

perhaps a new strain of a known bacterium, which initially received the name *Clostridium acetobutylicum Weizmann*. Unlike Fernbach's bacterium, which was capable of fermenting only potato starch, Weizmann's organism could digest starches from a wide range of vegetable sources, of which corn (maize) seemed the most promising.[53] Furthermore, butanol was a primary fermentation product. Acetone was another by-product that Weizmann isolated from the process in his laboratory; at the time this was of little interest, but when World War I broke out in August 1914, Britain found itself severely short of acetone, which was an essential ingredient in the manufacture of cordite, used as a propellant for explosive shells and thought to improve their quality.[54] This led to the idea that the "Weizmann organism," as *Clostridium acetobutylicum Weizmann* had come to be called, could be used to produce acetone on an industrial scale. To examine the feasibility of such a process, Weizmann was contacted by central figures in the British explosives industry and those involved in supplying these materials to the Admiralty.[55] In 1915, as noted above, Weizmann registered his patent on the process.

Weizmann's new independent project see Weizmann to Simon, *CWL*, VI, no. 83 (June 14, 1913). On funding, see Weizmann to Simon, *ibid.*, no. 90 (June 20, 1913).

53 Weizmann found the microbe on corn. That he searched for the microorganism he needed on the plant best suited to be used as a raw material demonstrates a biological mode of thinking. This, and the laboratory procedures he developed in its wake, would continue to guide Weizmann for many years to come. A 1938 article he published in the *Journal of the Society of Chemical Industry* puts his method in a nutshell, in the form of a lesson for beginners. [See no. 53 in the list of Weizmann's publications in this volume. – Ed.]

54 Cordite shells produced little smoke and thus made it difficult for the enemy to identify the source of fire. See D.H. Killeffer, "Butanol and Acetone from Corn: A Description of the Fermentation Process," *Industrial and Engineering Chemistry*, 19 (1927), pp. 46–50.

55 Weizmann to Frederick Nathan, *CWL*, VII, no. 157 (April 25, 1915). Nathan at this time served as adviser to the Admiralty on cordite supplies. See also Weizmann to William Rintoul, *ibid.*, no. 159 (May 2, 1915). Rintoul was chief research chemist of Nobel Industries Ltd., the large explosives factory in Ardeer on the coast at

When Weizmann demonstrated his discovery in his laboratory in Manchester, where it was observed and confirmed by university experts, it was seen as an achievement for the entire university and an example of how its scientists were contributing to the war effort.[56] What was now needed was to adapt the process to an industrial production scale.[57] At first, Weizmann thought that this could be accomplished while he continued to live and work in Manchester, but it soon became apparent that he would have to move to London. There he was provided with facilities at the Lister Institute for Preventive Medicine in Chelsea, the British equivalent of France's Pasteur Institute. Weizmann was now employed by the British Admiralty and found himself involved in activities in a field with which he had no previous experience.

The scaling-up process had two stages. First, the J & W Nicholson & Co. gin distillery in East London was requisitioned and used to prove

Ayrshire, Scotland. See *Trial and Error*, pp. 218–222, which also notes Speakman's role in transferring the acetone production technology to Canada.

56 See G.N. Burkhardt, "Weizmann and the University of Manchester," CWA, 65-20–79 (101). This two-page typed document apparently summarizes a lecture given by its author, who was a member of the Chemistry Department at the time. He cites a report, signed by Professor H.B. Dixon on behalf of the department, indicating that university chemists had scrutinized the "Weizmann process" and recommended it to the admiralty: "In the bio-chemical laboratories Dr. Weizmann had worked out a method of fermenting starch into butyl alcohol and acetone. The latter product being required for the manufacture of cordite and similar smokeless powders. … Dr Weizmann improved the process, and at my suggestion it was tested by chemists … and offered to the Admiralty." The university then freed Weizmann to work in London at the Lister Institute. Photocopies of this report can be found in CWA, Manchester documents. Biographies of Weizmann tell the story differently, indicating that Weizmann's contacts with people outside the university, such as Charles P. Scott, editor of the *Manchester Guardian*, and Sir Frederick Nathan of the Office of the Admiralty, helped him bring his discovery to the attention of British government figures; see, e.g., Reinharz, *CW*, II, pp. 40–72; and Norman A. Rose, *Chaim Weizmann: A Biography*, New York: Viking, 1986. The documentation cited here, however, demonstrates that the University of Manchester was instrumental in making the connection between Weizmann's laboratory, the government, and the munitions industry.

57 See *Trial and Error*, pp. 218–222.

that acetone could indeed be produced in industrial quantities. However, this facility allowed only limited production, and so the process was transferred to the British navy's cordite factory at Holton Heath in Dorset and other locations in Great Britain.[58] Weizmann also found himself training operators for the manufacturing process. When it became apparent that Britain lacked sufficient vegetable raw material, production was transferred overseas to facilities with fermentation and distillation equipment: liquor distilleries in Toronto and elsewhere in Canada, where agricultural produce was abundant. Weizmann's students and research assistants helped with the technology transfer and its implementation.[59]

Weizmann thus became not only the first to produce acetone in the laboratory using bacterial products instead of chemical reagents, but also the first to propose a bacteriological process that could be implemented on an industrial scale. Eventually, his research served as the starting point for two developments in which he himself did not take part. The

58 See the Wikipedia article on the Holton Heath factory, in which Weizmann is given considerable credit: https://en.wikipedia.org/wiki/Royal_Navy_Cordite_Factory,_Holton_Heath. He is also mentioned in the short documentary on the factory's history and capacity, www.greenacre.info/RNCF/ (both accessed November 29, 2016).

59 A report on the factory in Toronto tells of the technology transfer and the pride of those involved in the successful establishment of the factory there. See Col. A.E. Gooderham, "Report of the British Acetones Toronto, Limited, Toronto, Canada, May 1916–November 1918." Addressed to Sir Frederick Nathan, "Director of Propellant Supplies, London," it states: "I have the honor to submit herewith a report on the work done by the British Acetones Toronto, under the Weizmann process." The documents may be found in the John P. Robarts Research Library at the University of Toronto and on Toronto's Distillery District Heritage website: www.distilleryheritage.com/PDFs/documents/BAcontents.pdf. Speakman, who was involved, wrote the "Fermentation Department" chapter of this report. See also idem, "The Production of Acetone and Butyl Alcohol by a Bacteriological Process," *Journal of the Society of Chemical Industry*, 38 (1919), pp. 155–161. The production of acetone in this Toronto brewery is proudly trumpeted on the above site as a notable part of Canadian history: http://distilleryheritage.com/snippets/38.pdf (accessed November 29, 2016).

first was the biochemical investigation of the intermediate steps of the Weizmann process, as described, for example, in the successive early editions of the organic chemistry textbook written by Louis Fieser and Mary Fieser.[60] The second took place after the war ended: With acetone was no longer in such demand for military uses, the Weizmann organism offered a way of producing butanol, needed by the automobile and solvents industries for the lacquer paint used on cars.[61]

However, these developments were not connected to Weizmann's "biotechnological" vision and research policy, which he formulated only fifteen years later, with the founding of the Sieff Institute in Rehovot.[62] In the 1930s and 1940s, Weizmann was unsuccessful in promoting his idea of an industry based on the fermentation of agricultural products. Only after his death, with the growth of interest in biotechnology in the final two or three decades of the twentieth century, was Weizmann's work recognized as groundbreaking. Although he did not see himself as engaged in "biotechnology" (a term that did not come into use until the 1970s), researchers and historians of the field view him as a pioneer in the field.[63]

60 Louis F. Fieser & Mary Fieser, *Organic Chemistry*, Boston: D.C. Heath, 1944, pp. 457–462. This constitutes early evidence of the integration of bacteriological knowledge in an organic chemistry textbook for beginners.

61 A series of articles on the Weizmann process appeared in *Industrial Engineering Chemistry* at the end of the 1920s; see, e.g., C.L. Gabriel, "Butanol Fermentation Process," *Industrial and Engineering Chemistry*, 20 (1928), pp. 1063–1067, http://pubs.acs.org/doi/pdf/10.1021/ie50226a020 (accessed November 29, 2016). The financial sources that provided Weizmann with economic independence and prosperity have not been examined, but it seems reasonable to assume that a considerable portion derived from royalties paid him for the widespread use of his discovery in these industries.

62 See Chapter III of this article.

63 For scientists, see, e.g., Jones & Woods (above, note 35). See also Mark M. Green & Harold A. Wittcoff, *Organic Chemistry Principles and Industrial Practices*, Weinheim: WILEY-VCH, 2003, pp. 223–224. For historians of the field, see Bud, *Uses of Life* (above, note 30), esp. the chapter entitled "The Weizmann Process" (pp. 37–45). Bud stresses the connection between Fernbach and Weizmann. See also Anthony S. Travis, "Chaim Weizmann: Chemist, Biochemist, and Father of

In any case, Weizmann's shift to the acetone project, for which he paid the price of leaving the University of Manchester, must certainly at first have seemed to him beneficial for his Zionist interests. London was a center of Zionist activity and a place that afforded Zionist movement leaders contact with public and government figures. These were all critical to the achievement of the Balfour Declaration and to subsequent developments arising from the declaration and from the British conquest of Palestine from the Ottomans. Weizmann's increased involvement in these affairs toward the end of the war and after was detrimental to what he saw as the "organic" integration of his science and his Zionism. He would not return to the University of Manchester until 1919, when the university granted him an honorary degree.[64]

It was to no small measure Weizmann's involvement in the acetone project that gave him standing and enhanced his contact with British elites. When southern Palestine came under British control and the Balfour Declaration was issued, Weizmann was appointed to head the Zionist Commission, the Zionist body assigned the task of helping to implement the Balfour Declaration. Weizmann now found himself responsible for the Zionist Organization's academic policy and for the execution of one of its central goals – the establishment of the Hebrew University of Jerusalem.

II. The University Plans, 1902–1925

Beginning in 1902, Weizmann was involved in planning a Jewish institution of higher education and, later, a Hebrew university and research institutes. The various incarnations assumed by these plans occupied

Biotechnology," *The Biochemist* (Aug.–Sept. 1992), pp. 15–19. Weizmann's pioneering status is also highlighted by his discovery's inclusion on the list of "Significant Events in Microbiology 1861–1999" (above, note 46).

64 LLD *honoris causa* (1919); see www.manchesterjewishstudies.org/weizmann-manchester-university (accessed November 29, 2016), and see Burkhardt, "Weizmann and the University" (above, note 56).

him for much of his life, reflecting his ever-widening acquaintance with academic institutions and research organizations of different types. These included a polytechnic and a technological university in Germany,[65] universities in Switzerland, and the University of Manchester in England, as well as medical research organizations such as the Pasteur Institute in Paris, and, of course, industrial and military research laboratories.

The changes in Weizmann's academic and research policies – and the two should be differentiated as distinct components of his science policy – mirrored the progress of his academic standing and position in the world of science. They also reflected his progressively more varied research experience and his personal acquaintance with members of the scientific elite and leaders in the field of science policy at the time, first in Europe and later in the United States. Interwoven with all these was his Zionist ardor to establish his own research laboratory, in his own specialization, in Palestine.

The contents of Weizmann's plans all shared a number of basic components, but these received different emphases over time.[66] The new institution was needed to meet the demand for higher education among Jewish students who had difficulty gaining acceptance into European universities (including higher technological institutes), and to provide an academic home for established Jewish scientists and scholars who had been rejected by European institutions. Beyond this, the Jewish university would provide an institutional framework for scientific research and foster a renaissance of Hebrew culture, as well as providing material infrastructure for the Zionist settlement enterprise through medical, agricultural, and industrial research.[67] But Weizmann's vision

65 Before coming to Berlin, Weizmann had studied for about a year at the Technische Universität Darmstadt. See Reinharz, *CW*, I, p. 29.

66 Weizmann wanted to be involved in the plans for a Technion in Haifa when they were first drawn up, but only for personal reasons: He was interested in an academic post and a laboratory there. See Reinharz, *ibid.* The present chapter examines those of Weizmann's programs that had more operational dimensions.

67 The connection between universities and nationalism is a multifaceted phenomenon. For examples of universities that were founded and flourished in connection

and plans included another element that took on ever more importance as time went by – the promotion of scientific excellence in pure research as an eminent, universal cultural value in its own right. "We have made up our minds," he declared in 1925, "that it is for our University to win its spurs and build up its reputation by the distinctive value of its contributions to the common stock of knowledge."[68] As Weizmann's

with nation-building, see John E. Craig, *Scholarship and Nation Building: The Universities of Strasbourg and Alsatian Society, 1870–1939*, Chicago: University of Chicago Press, 1984; and J. Gwynn Williams, *The University College of North Wales: Foundations 1884–1927*, Cardiff: University of Wales Press, 1985. For a brief survey of the connection between nationalism and science, see Paul T. Durbin (ed.), *A Guide to the Culture of Science, Technology and Medicine*, New York–London: The Free Press, 1980, pp. 31–33. Ronald C. Tobey distinguishes between central management of scientific research harnessed to national needs – for example, in wartime – and the scientific system's influence on establishing scientific values and integrating them into the cultural values of the surrounding society. See idem, *The American Ideology of National Science, 1919–1930*, Pittsburgh: University of Pittsburgh Press, 1971, pp. 3–61. For Weizmann and the Zionist movement, these two aspects seem closely linked. The struggle to establish a dominant national language at national universities is mentioned frequently in Rüegg's *History of the University in Europe* (above, note 2). With regard to the Hebrew University of Jerusalem, the responsibility for promoting the national language is symbolically underscored by the institution's very name, and the nature of the research that was to be performed there and the dissemination of its results took up an important part of the plans for its establishment. See "The Revival of Hebrew," in *Inauguration, April 1, 1925 – The Hebrew University of Jerusalem*, Jerusalem: Hamadpis Press, [Nisan/April] 5685/1925 (henceforth: *Inauguration*, Large Format), pp. 17–20. This was a large-format booklet in three languages, Hebrew, Arabic, and English, which apparently was handed out to the ceremony's attendees. Another booklet with the same title, in a smaller format and in Hebrew and English only, was printed later the same year, in Heshvan/October–November 5686/1925 (henceforth: *Inauguration*, Small Format).

68 *Inauguration*, Small Format (above, note 67), p. 23. Compare this with Weizmann's 1914 statement: "More than anything else in the universe, the university's most important scientific and cultural purpose is, in my view, to be a political force by means of underlining the universal and eternal foundation of our cultural enterprise." "Address to Jewish Students, Heidelberg, 10 Sivan 5674 [June 4, 1914],"

commitment to pure research grew, so did his enthusiasm for an institution standing at the forefront of research; concomitantly, his interest in a Jewish university for the academic training of young Jews waned.

The Plan for a Jewish Institute for Higher Studies, 1902

In 1902, on the initiative of the Zionist movement's Democratic Fraction, which stressed the promotion of a Jewish cultural revival, Weizmann, together with Martin Buber and Berthold Feiwel, published *Eine jüdische Hochschule*, a pamphlet presenting the first version of the plan for a Jewish academic institution.[69] The plan was structured around quantitative information, heralding the new data-based discourse characteristic of Jewish scholars at the end of the nineteenth century. Writers used the social sciences, supplemented by statistics, to arrive at definitions of Jewish identity. Analyses of this sort provided the foundation for the political arguments and practical programs that

in *Chaim Weizmann on the Hebrew University*, Jerusalem: General Committee of the Friends of the University in Eretz Israel, 1945, p. 20 (Hebrew).

69 Martin Buber, Berthold Feiwel & Dr. Ch. Weizmann, *Eine jüdische Hochschule*, Berlin: Jüdischer Verlag, 1902. In 1968, Magnes Press published a facsimile of this German-language pamphlet with a Hebrew translation by Shaul Esh on facing pages. There are slight differences between the Magnes version and the copy of the 1902 second edition held in the library of the Johann Wolfgang von Goethe-Universität in Frankfurt, which can be viewed online at http://sammlungen.ub.uni-frankfurt.de/freimann/content/titleinfo/1141634 (accessed December 5, 2016). I am grateful to Alexander Elon for referring me to this source. The pamphlet's motto, "Give me Yavneh and its sages," is taken from rabbinic legend: After Rabban Yohanan Ben-Zakkai fled the besieged and soon-to-be-destroyed Jerusalem, he entreated the Roman military commander Vespasian (later the Emperor) to spare the yeshiva at Yavneh so that Jewish learning could continue through the generations. The epigraph testifies to the authors' commitment to the vision of a spiritual center for Jews throughout the world. On the pamphlet as a Democratic Fraction initiative, see *Trial and Error*, pp. 91–94 and elsewhere; for a discussion, see Reinharz, *CW*, I, pp. 130–132; Israel Kolatt, "On The Idea of the Hebrew University in the Jewish National Movement," in Shaul Katz and Michael Heyd (eds.), *The History of the Hebrew University of Jerusalem*, I: *Origins and Beginnings*, Jerusalem: Magnes, 1997, pp. 73–74 (Hebrew).

formed the basis of Jewish national hopes.[70] This characteristic reliance on observations and measurements, data and comparisons, would be evident in Weizmann's later plans as well.[71]

70 I am grateful to Ari Barel for referring me to Mitchell B. Hart, *Social Science and the Politics of Modern Jewish Identity*, Stanford: Stanford University Press, 2000; on the connection between the Jewish statistical project and the Democratic Fraction, see pp. 36–39. The Zionists' faith in the social sciences and statistics as a means of substantiating the notion of a Zionist national liberation is discussed in a number of places, e.g., pp. 53–55, 214–216, and elsewhere. Oren Soffer addressed the Jewish nationalist interest in statistics in *There Is No Place for Pilpul! The Periodical Hatzefirah and the Modernization of Sociopolitical Discourse*, Jerusalem: Bialik Institute–Hebrew University, 2007 (Hebrew), e.g., pp. 118–120. However, the forums involved in planning the Zionist agricultural settlements in Palestine seem to have preceded those of the social sciences in basing their arguments on systematic surveys and quantitative data, though without scientific pretensions. An example can be seen in the records of the Society for the Support of Jewish Farmers and Artisans in Syria and Palestine (the Odessa Committee) for the year 5660 (1899/1900). After visiting Jewish farming colonies in Palestine, Asher Ginzberg (Ahad Ha'am) and agronomist Avraham Zusman provided the committee with quantitative and comparative findings. See *Critique: The Jewish Colonies in the Holy Land in 5660* [1900], http://web.nli.org.il/sites/NLI/Hebrew/library/Pages/BookReader.aspx?pid=2688929 (Hebrew; accessed December 5, 2016).

71 The data-based discourse in the pamphlet by Buber, Feiwel, and Weizmann (above, note 69) in and of itself clearly disqualifies it from being the final link in the long series of earlier proposals for the establishment of a Jewish university. In particular, it should *not* be seen as a continuation of the plan enunciated by Zvi Hermann Schapira (1840–1898) for a "Beit Midrash Taḥkemoni" in what would become Jewish Palestine. Although Weizmann remarked in 1913 that Schapira had supported the establishment of a Jewish university, a careful reading of Schapira's proposal shows that he was thinking of a very different sort of institution. At the First Zionist Congress in 1897, he declared that his proposal to establish an institute of higher learning in Eretz Israel did not overtly refer to a university. He envisaged a "special center" for the cultivation of Jewish spiritual, religious, and moral education. With one joint department and three specialized sub-departments – theology, theoretical science, and technical and agronomic studies – the institution would offer an integrated curriculum of religious, scientific, and practical studies. Schapira's scheme eschewed the standard division between gymnasium and university and aimed instead to prepare its students for independent study and work in the sciences and in practical activities. See Haim Orlan (ed. and transl.),

In the 1902 program, published while Weizmann was still a *Privatdozent* at the University of Geneva, the *jüdische Hochschule* is presented as a combination of "a university (on the German or Swiss model) and a polytechnic."[72] In other words, in today's terminology, it would be a multi-disciplinary teaching university. Its purpose would be to train Jewish youth in higher disciplines, with special attention to Jewish studies, as well as in technical professions.[73] In other words, it would be a teaching university,[74] but one that incorporated elements of a polytechnic. On the research side, it followed German models whereby each professor was responsible for obtaining outside funding to equip and maintain his laboratory and to support his research projects. The exception in this case was Jewish studies, where guaranteeing "free scholarly inquiry" was explicitly made part of the university's purview.[75]

The pamphlet set out the reasons for establishing a Jewish institute of higher education: to further the national cause; to respond to the plight both of young Jews and of prominent Jewish intellectuals who, because of anti-Jewish prejudice, had trouble enrolling or finding teaching and research positions in European, especially Russian, universities, and

Minutes of the First Zionist Congress in Basel, 1–3 Elul 5657/29–31 August 1897, Jerusalem: Reuben Mass, 1997, pp. 160–161 (Hebrew). In nearly every work on the history of the university and of science and scholarship in Israel, Schapira is presented as the great visionary who anticipated the Hebrew University of Jerusalem. He was recently even commemorated on the Founders Walls at the entrances to the Hebrew University campuses on Mount Scopus and Giv'at Ram. How he earned this honor, given his patent refutation of the idea of a traditional university, is an interesting question that lies outside the scope of the present study.

72 *Eine jüdische Hochschule* (above, note 69), p. 32.

73 *Ibid.*, p. 31. It also provided for general studies, philosophy, political science, pedagogy, Jewish studies, mathematics, and natural sciences, to be supplemented by the training of architects and civil, mechanical, and electrical engineers. In addition, it would encompass a technical chemistry school and a school of forestry and agriculture; see *ibid.*, p. 32.

74 The model was explicitly German, as can be seen from the composition of the proposed faculty: 60 salaried professors, 25 teaching assistants, 50 *Privatdozenten*, 20 administrative employees, maintenance staff, etc.; see *ibid.*, p. 35.

75 *Ibid.*, p. 20.

"do not want to turn their backs on their people" (that is, convert to Christianity); and also to serve as a symbol of the creative Jewish spirit and a spiritual and intellectual center (*geistiges Zentrum*) for all who wished to see the Jewish people endure and thrive.[76] If the university were built in Palestine, it would not only establish a moral claim for ownership of the land, but also contribute to "the country's technical development and the people's education from an economic point of view."[77]

The Plan for a Hebrew University in Jerusalem, 1913

At the Eleventh Zionist Congress, convened in Vienna in 1913, Weizmann envisioned a plan for a university radically different from the one he had projected together with Buber and Feiwel in 1902. As he now described it, the new institute would be a Jewish national research university located in Jerusalem, whose national character would be expressed in its name, its language, its location, and its role as the premier center of Jewish culture.[78] The 1913 plan also differed notably from the 1902 plan with regard to the academic nature of the institution, in that it did not include polytechnic fields.[79]

76 *Ibid.*, pp. 19–20.

77 *Ibid.*, p. 21.

78 *Stenographisches Protokoll der Verhandlungen des XI. Zionisten-Kongresses in Wien, vom 2. bis 9. September 1913*, Berlin–Leipzig: Jüdischer Verlag, 1914, pp. 300–309; *Weizmann on the Hebrew University* (above, note 68), p. 13. Weizmann spoke about a Jewish university, but when the chairman, Moses Gaster, introduced him, he announced that Weizmann would lecture on "die hebräische Universität in Jerusalem" (*Stenographisches Protokoll*, p. 300). The Congress's resolutions also referred to the "Hebrew" university; *ibid.*, p. 368. In his 1920 lecture "A Sanctuary of Glory for the Jewish Spirit," Weizmann renounced the view he had expressed in 1913 on the gradual implementation of the Hebrew language in the incipient university; *Weizmann on the Hebrew University* (above, note 68), p. 31.

79 It seems reasonable to suppose that the shift from a *jüdische Hochschule*, which would be a combination of "a university (on the German or Swiss model) and polytechnic," to a vision of a university alone resulted from the ongoing establishment of the Technion in Haifa at this time. Weizmann sought involvement in the plans

The new plan still sought to provide an academic home for aspiring Jewish students and scholars who could not find places in Europe. However, in order to persuade young Jews that a degree from the university in Jerusalem would be on a par with those of European institutions, it proposed to recruit leading figures in the relevant fields of instruction. Where would they be found? Weizmann's speech at the Eleventh Zionist Congress concisely lays out a central element of his academic policy: "Efforts should be made to enroll young, highly promising workers .... Teachers must at the same time be good research workers, and free research [*freie Forschung*] must be allowed its proper place."[80] Faculty members would thus have a mandate to conduct research, and this would be among the institution's concerns.

The primary function of the university would be to train professionals and to disseminate scholarly knowledge to the population of Palestine and all the lands of the [Middle] East.[81] At the same time, however, Weizmann emphasized the university's role as "an intellectual center where Jews could teach, learn and research in a favorable atmosphere."[82] In other words, the concept of a multi-disciplinary teaching institution was replaced by that of a research university.

Weizmann based the feasibility of establishing a university in Jerusalem on several historical examples of the founding of universities. He cited data on finances, operation, and personnel from the medical faculty of the American University of Beirut, which had opened in the late nineteenth century, and from the Université Libre de Bruxelles, "among the finest universities of Europe."[83] Alongside these, apparently to make a better case for implementing his plan, Weizmann noted that an institute of bacteriology and pathology was already in forma-

for the Technion as well; see Reinharz, *CW*, I, pp. 338–340 and elsewhere. But this matter requires separate attention.

80 "A Jewish University for Jerusalem," Speech at the Eleventh Zionist Congress, *CWP*, no. 21, p. 107. The full text of Weizmann's address is on pp. 101–113. For the German text, see *Stenographisches Protokoll* (above, note 78), p. 306.

81 *CWP*, no. 21, *loc. cit.* (above, note 80).

82 *Ibid.*, p. 102.

83 *Ibid.*, p. 107.

tion in Jerusalem and, he said, could serve as the nucleus of a medical school,[84] which had not been part of the 1902 program.[85] Weizmann reported that a budget for a medical faculty had already been drawn up and he believed that establishing a faculty of law and political science would be simpler.[86]

However, Weizmann also offered an alternative to establishing these faculties, and it deserves special attention: "A closer examination may prove that it would be preferable not to establish separate faculties simultaneously, but rather individual institutes. Thus it has already been suggested that we establish a Chemical Institute as a necessary addition to the Serological Institute now being founded … as well as … the existing experimental station."[87] As will become clear below, the establishment of research institutes would indeed become Weizmann's preferred path to founding a university in Jerusalem. As early as 1912, he believed that he could best achieve his goal via a modular process, based on research institutes that already existed or would soon be es-

84 *Ibid.*, p. 108.

85 *Eine jüdische Hochschule* (above, note 69), p. 32.

86 *CWP*, no. 21, p. 108.

87 *Ibid.* The "Serological Institute" was the Pasteur Institute in Jerusalem, founded by Dr. Aryeh Boehm; see idem, "The History of the Pasteur Institute During its Fifteen Years of Existence," in *The Pasteur Institute in (Palestine) for Health, Medicine and Biology*, 1 (Tishrei 5689 [1929]), pp. 29–39 (Hebrew). For more on the Pasteur Institute in Jerusalem, see Nadav Davidovitch and Rakefet Zalashik, "Pasteur in Palestine: The Politics of the Laboratory," *Science in Context*, 23, Special Issue 4 (December 2010), pp. 393–399, www.ncbi.nlm.nih.gov/pubmed/21560596 (accessed December 6, 2016). The Experimental Station was the Jewish Agricultural Experimental Station run by Aaron Aaronson in Atlit. As for the bacteriological and pathological institute, Weizmann was probably referring to the Hebrew Health Station established by philanthropist Nathan Strauss, whose staff included a bacteriologist, Dr. Aryeh Goldberg, and an ophthalmologist, Dr. Aryeh Feigenbaum. See Shaul Katz, "Aaron Aaronson and the Beginning of Scientific and Agricultural Research in Eretz Israel," *Cathedra*, 3 (February 1977), pp. 3–29, especially Table 2 and its notes on p. 7 (Hebrew). These institutes were not yet in operation when Weizmann visited Palestine in 1907, and in retrospect it would seem that Weizmann, who was quite familiar with the Pasteur Institute in Paris, overstated the laboratories' research capabilities.

tablished, including the Technion.[88] Nevertheless, the 1913 plan did not present the fostering of research as a national goal in its own right. The most salient difference between it and the 1902 program was that the latter had been promoted only by a group of young members of the Democratic Fraction of the Zionist movement, while the former was adopted by the Zionist Congress. Thus, in 1913, the establishment of the Hebrew University in Jerusalem became a central component of the Zionist vision,[89] with Weizmann as its principal advocate. His association with the university project also helped establish his standing in the Zionist movement.

From the University Plan to a Plan for a Medical and Natural Sciences Research Institute in Jerusalem, 1914

In 1902, Weizmann and his two colleagues had hoped to fund their plan from public donations. In 1913, Weizmann preferred to seek the backing of a few affluent Jewish philanthropists. After the Eleventh Zionist Congress adopted his plan, Weizmann sent the Zionist Smaller Actions Committee a working program for setting up the university and its faculties.[90] It was understood that, to this end, Weizmann would meet with Baron Edmond de Rothschild in Paris. Upon making some preliminary inquiries, Weizmann learned that Rothschild did not support the establishment of a university but was prepared to help finance a research institute in Jerusalem that could perhaps eventually become a university. Weizmann reported that Rothschild was interested in something along the lines of the Rockefeller Institute or the Pasteur Institute, where senior scientists could conduct research, publish it from Jerusalem, and gradually attract students, ultimately culminating in the formation of a university. However, Rothschild wanted to get the opinion of medical researcher Paul Ehrlich (1854–1915, Nobel laureate 1908), director of

88 See Weizmann to Magnes, *CWL*, V, no. 278 (May 5, 1912).

89 *Stenographisches Protokoll* (above, note 78), p. 368.

90 Weizmann to the Zionist Smaller Actions Committee, *CWL*, VI, no. 156 (November 15, 1913).

the Königliches Institut für experimentelle Therapie zu Frankfurt (today, the Paul-Ehrlich-Institut), whom he greatly respected.[91]

The proposal to establish a research institute in Jerusalem certainly was not detrimental to Weizmann's plans. He had, after all, already suggested this as a possibility. Furthermore, many of his Zionist colleagues were prepared, for the time being, to forgo teaching in the university and accept Rothschild's condition, if this would lead to the establishment of a Jewish research institute in Jerusalem.[92] To pique Rothschild's interest, Weizmann, with Ehrlich's assistance, prepared a memorandum that differed radically from his previous proposal.[93] It called for the establishment in Jerusalem of a "jüdisches medizinisch-naturwissenschaftliches Institut," which would serve as an umbrella for a group of research institutes dealing, alongside medical research – physiology, bacteriology, and serology – with experimental natural sciences relevant to medicine, including inorganic chemistry, organic and biological chemistry, and agricultural chemistry, as well as experimental physics and physical chemistry.[94] These fields were not part of the Pasteur

91 Weizmann to Magnes, *CWL*, VI, no. 192 (January 12, 1914).

92 In a letter to Harry Sacher, Weizmann related that Nahum Sokolow, secretary-general of the World Zionist Congress, after learning of Rothschild's opinion, said: "If the baron would think, that one has to start with an astronomical Observatory, we would have to take his view." *CWL*, VI, no. 201 (January 17, 1914).

93 See "Das Königliche Institut für experimentelle Therapie zu Frankfurt a.M.," 1907, www.pei.de/SharedDocs/Downloads/institut/veroeffentlichungen-von-paul-ehrlich/1906-1914/1907-koenigl-institut-exp-therapie.pdf?__blob=publicationFile&v=1 (accessed December 6, 2016).

94 A memorandum inaccurately entitled "The University's Preliminary Structure" can be found in *CWP*, no. 23, pp. 113–118. The Weizmann Archive preserves an earlier version of this document on which Leopold Landau's signature appears alongside those of Ehrlich and Weizmann. Landau, a well-known German gynecologist with whom Weizmann was acquainted from his studies in Berlin, had a son, Edmund Landau (see below), who was married to Ehrlich's daughter. Leopold Landau persuaded his son's illustrious father-in-law to receive Weizmann. See *Trial and Error*, pp. 179–180. A version of the memorandum can also be found in the Central Zionist Archives, Z3/525; appended to it is a cover letter signed by Leopold Landau alongside Ehrlich and Weizmann, dated March 24, 1914, at Frankfurt

Institute's program, but Ehrlich's institute in Frankfurt had a physical chemistry laboratory, and including them in the Jerusalem institute would promote local independent research capabilities. The institute would also include a twenty-bed patient ward, presumably for experimental purposes. The original Pasteur Institute, founded in 1900, included a medical isolation ward.[95]

Weizmann was optimistic about the future of the institute, which was to be staffed with elite researchers. He hoped to recruit young and promising scientists by encouraging senior figures such as Paul Ehrlich and Edmund Landau to recommend to junior colleagues and promising students that they pursue their careers in Jerusalem. Weizmann believed that young scientists at the outset of their careers would be eager to work diligently at the institute in Jerusalem in order to make names for themselves. Departments would be headed by scientists with already established reputations, such as Alexandre Besredka, an immunologist at the Pasteur Institute, and physicist Leonard Ornstein of the University of Utrecht. Weizmann raised the possibility that he himself might head the chemistry department. He hoped prominent Jewish scholars who could not move permanently to Jerusalem would come to teach short courses or work there for a time,[96] and that ongoing ties could be established with highly regarded research institutes in Europe and the United States. He even dared raise the possibility, which, he declared, was not utopian, of a member of the staff winning a Nobel Prize.[97]

The plan to set up a Jewish institute of medicine and natural sciences in Jerusalem was cut short by the outbreak of war in August 1914. Nevertheless, its very proposal shows that, as far as Weizmann was concerned, establishing an institute focusing on medical research in its

am Main. See Shifra Kolatt, "Introduction," *CWL*, VI. See also Reinharz, *CW*, I, pp. 377–378, 393–394, and elsewhere.

95 See Delaunay, *L'Institut Pasteur* (above, note 50). No mention is made of a chemistry or physics department.

96 Weizmann to Magnes, *CWL*, VI, no. 350 (May 25, 1914). The budget for the proposed institute included a sizable entry for travel allowances.

97 Weizmann to the Zionist Smaller Actions Committee, *CWL*, VI, no. 310 (April 26, 1914).

broadest sense, without offering instruction, could also be considered an important Zionist goal – the more so if it were to include a department of biological chemistry, that is, one that he himself could head. Ehrlich's participation in the preparation of the memorandum presenting the proposal and his consent to head the institution's academic committee constituted Weizmann's first successful enlistment of a leading scientist in support of this Zionist academic project. Later successes of this sort were the recruitment of Albert Einstein (1879–1955, Nobel laureate 1921) in support of the Hebrew University project[98] and of Richard Willstätter (1872–1942, Nobel laureate 1915) for the Sieff Institute (see below). However, the roles played by Einstein and Willstätter were quite distinct from the assistance that Ehrlich lent Weizmann. As noted, Weizmann had already made the acquaintance and gained the friendship of noted scientists, but Ehrlich was different in that he had already set up and operated a research institute and was willing to share his experience and knowledge with Weizmann. It would not be far-fetched, then, to see Ehrlich as Weizmann's mentor in taking the appropriate steps to establish a center of scientific excellence.[99]

The institution proposed in the 1914 plan never received a name in Hebrew, nor any mention in most histories of the Hebrew University. Yet the main elements of Weizmann's science policy, most centrally the goal of scientific excellence, were evident in it and would recur in his

98 On Einstein and the Hebrew University, see Ze'ev Rosenkranz, "'Lofty Spiritual Centre' or 'Bug-Infested House?' Albert Einstein's Involvement in the Affairs of the Hebrew University, 1919–1935," in Katz and Heyd, *History of the Hebrew University*, I (above, note 69), pp. 386–394; Ze'ev Rosenkranz, *Einstein before Israel: Zionist Icon or Iconoclast?*, Princeton: Princeton University Press, 2011, Chap. 3, pp. 99–138 and elsewhere. See also Walter Isaacson, *Einstein: His Life and Universe*, New York: Simon & Schuster, 2007, Chap. 13; Hanoch Gutfreund, "Albert Einstein: The Man, the Jew, and the Zionist," *Scientific American Israel* (January 2005), http://lib.cet.ac.il/pages/item.asp?item=11349 (Hebrew; accessed December 6, 2016).

99 In later years, the idea of founding research institutes as a first step toward a university would be attributed to Ehrlich, in his role as consultant to the Zionist Organization's University Committee. See the unsigned chapter "The University," in *Keren Ha-Yesod Book*, London: Leonard Parsons, 1921, p. 137.

later enterprises, from 1920 onward, to found the Hebrew University of Jerusalem, the Sieff Institute (1933) and the Weizmann Institute of Science (1944–1946). In fact, the two latter institutions fully embodied the research institute model that made its first appearance in the 1914 plan.[100]

The "Complete University" Plan, 1918

After the Balfour Declaration and the British conquest of southern Palestine and Jerusalem, the Zionist Commission headed by Weizmann arrived in the newly conquered areas. Among the commission's explicit goals was to further the foundation of a university in Jerusalem. After negotiations, the British military authorities granted permission for a cornerstone-laying ceremony on the fifteenth of the Hebrew month of Av, July 24, 1918. In fact, more than a dozen "cornerstones" were laid.[101] In the address that Weizmann gave in English at the ceremony on Mount Scopus, and in a Hebrew-language article published a short time later in *On the Threshold*, a collection issued by the Po'ale Zion party, he repeated the main points of the 1913 plan regarding both the institution's academic and organizational structure and its purpose of providing an academic home for young Jews eager to pursue higher studies.[102] He described the future university and its faculties:

100 Compare the plan for the operational aspects of the Sieff Institute and its research programs, signed by Weizmann: "Scheme for Daniel Sieff Research Institute," London, December 15, 1933, *CWP*, II, no. 8, pp. 33–38.

101 "Our Festive Day: Details of the Cornerstone Laying for the Building of the Hebrew University in Jerusalem, 15 Menahem-Av 5678 [July 24, 1918]," Jerusalem: Azriel Press, 1918 (Hebrew). Benjamin Z. Kedar points out that, despite the title given the ceremony, several "foundation stones" were laid rather than a single one. See Kedar, "Laying the Foundation Stones of the Hebrew University of Jerusalem, July 24, 1918," in Katz and Heyd, *History of the Hebrew University*, I (above, note 69), pp. 90–119.

102 For the speech, see Kedar, "Laying the Foundation" (above, note 101), pp. 110–119; for the Hebrew article, see Chaim Weizmann, "On the Hebrew University," in *On the Threshold: Collection on Life and Literature*, Jerusalem: Po'ale Zion Party, 1918, pp. 16–20 (Hebrew); reprinted in *Weizmann on the Hebrew University* (above, note 68), pp. 21–24.

"a faculty of medical studies with all the requisite departments and adjunct institutions: laboratories and clinics"; adjacent to it would be "a special faculty for natural science" and "a department of philosophical studies, or if you wish to call it by another name, of Jewish studies, a kind of *Faculté de Lettres*, which would include, according to today's parlance, the humanities, social sciences, and most importantly Jewish studies."[103]

The envisaged university would not only include the general fields of study and research typical of European universities, but would also assign a central place to Jewish studies and the Hebrew language, in response to the demand for research on the Jewish national heritage and, in particular, the Jewish textual heritage. Furthermore, Weizmann relates to the specific locale to which the university would provide a response: "from the outset of the Hebrew University, an important place must be assigned to the establishment of an agricultural experimental station … which can easily be expanded and turned into a special Department of Agricultural Studies."[104]

Weizmann also stressed that certain subjects were especially appropriate for research at "our Hebrew University." One of these was archaeology, which had brought so much to light in Egypt and Greece and promised discoveries in Palestine as well. Others were chemistry, bacteriology, geology, and climatology, which would combine to increase the land's fertility. The country's two endemic diseases, malaria and trachoma, would be eradicated with the help of basic and applied research. In this way the university would contribute to the establishment of the national home.[105]

While the university would aspire to the "highest scientific level,"[106] it would also, as the 1913 plan had specified, disseminate professional knowledge and provide adult education.[107] In this way, "the Jewish

103 Weizmann, *On the Threshold* (above, note 102), pp. 16–20.

104 *Ibid.*, p. 19.

105 See the critical edition of Weizmann's speech in Kedar, "Laying the Foundation" (above, note 101), pp. 116–117; Weizmann, *On the Threshold*, (above, note 102).

106 Weizmann's speech in Kedar, "Laying the Foundation," p. 117.

107 *Ibid.*, p. 118.

workman and farm-labourer" would be able to complete their education in their spare hours.[108] At the time, the type of institution Weizmann was describing was called a "complete university." That it was nevertheless to be a research university is evident from his address, in which the nurturing of research was inextricable from the provision of academic and research opportunities to young Jews thirsty for them:

> No teaching can be fruitful nowadays, unless it is strengthened by a spirit of enquiry and research, and a modern University is not only to produce highly trained professional men, but to give ample opportunity to those capable of and ready to devote themselves to scientific research, to do so unhindered and undisturbed. Our University will thus become the home of those hundreds of talented young Jews, in whom the thirst for learning and critical enquiry have been ingrained by heredity throughout ages, and who in the great multitude of cases are at present compelled to satisfy their burning need amid un-Jewish, very often unfriendly surroundings.[109]

Back to the Plan for Research Institutes, 1919–1925

About a year after the Mount Scopus ceremony, the plans for a multi-faculty university were severely curtailed. In September 1919 the Zionist Organization office in London drew up a draft decision for submission to the University Committee, which echoed the 1914 plan to establish an umbrella institution for a small number of research institutes. Rather than establishing teaching faculties, which was put off indefinitely, the proximate goal was to set up research institutes in "physics, chemistry, biology, geology, mineralogy, agriculture." These would later serve as the nuclei for faculties of natural science and medicine.[110]

108 *Ibid.*

109 *Ibid.*, p. 113.

110 *Records of the Zionist Executive Committee*, I: *February 1919–January 1920*, ed. Yehoshua Freundlich and Gedalia Yogev, Tel Aviv: Tel Aviv University–Hakibbutz Hameuchad, 1975, pp. 229–230 (Hebrew). Regarding Jewish studies specifically

At the beginning of 1920 the plan was cut back even further, as attested to by a typed "Progress Report" – stenciled, indicating that it was intended for wide distribution[111] – stressing the commitment to the establishment of a superior academic research institution, while basing it, at least at its start, on only four research institutes: physics, chemistry, microbiology, and a "Hebrew Institute and Department of Jewish Studies." Two justifications are offered for this smaller plan. First, for the time being the founders did not have the means to construct the buildings required for classes. Second, the reduction in the number of fields was in line with academic policy, most certainly reflecting Weizmann's position:

> The guiding principle must be to emphasise the quality even at the expense of size and quantity. ... it was agreed that it would be most appropriate to start the Science Departments as Research Institutes and not as Teaching Faculties. Such institutes, even if small, could reach a high standard in spite of the limited funds ... they will ... establish the reputation of our University, thus ensuring it against the danger of a low scientific standard and a corresponding low

and the humanities in general, this document states: "The appropriate field for original research of general scientific value and intellectual activity of general Jewish importance at the Hebrew University in Palestine, will be found in the study of Hebrew and Semitic civilization and the study of the Near East. For this purpose we must establish an academy for Hebrew, Semitic, and Near Eastern studies, which will constitute part of the future humanities faculty." The extent to which this was meant to replace the proposal for a "Hebrew Institute and Department of Jewish studies" may be seen below. The "proto-Canaanite" spirit that this document seems to express is worthy of study in and of itself.

111 "Report: Preparations for the University in Jerusalem (to be Submitted to the Annual Conference of the Zionist Organization)" (Hebrew and English; henceforth: "Preparations Report"), stencil, London. The year is not noted, but the content indicates that it was written after March 1920; see National Library of Israel, 2° 34 2998. For a detailed discussion of the preparation of this report, see Hagit Lavsky, "From Foundation Stone to Opening: The Establishment of the Hebrew University, 1918–1925," in Katz and Heyd, *History of the Hebrew University*, I (above, note 69), pp. 136–143.

> reputation like that from which new universities in backward countries especially in the East generally suffer.[112]

In other words, the 1920 plan aimed at scientific excellence while temporarily deferring teaching. Previous versions of the plan, certainly that of 1914, had already aspired to quality research, but this was the first time that it became the central goal, with quality privileged over quantity. In the 1913 plan, scientific quality was cited as a way of drawing students to study for a degree that would be respected in Europe. In 1920, a high level of research was a goal in and of itself. No mention is made of relieving the hardships of Jewish students and scholars rejected by European institutions, nor of the establishment of a Jewish spiritual and intellectual center. Rather, the university is meant to participate in the international scientific enterprise by producing superlative scientific work, as befits a university representing the Jewish people.

There were two reasons to begin with research institutes. First, in the spirit of the guiding principles cited above, research institutes were the best way of earning the university a good name. Second, since the goal of the founders was, in future, to establish faculties of medicine, agriculture, and, of course, natural sciences, the proposed research institutes were appropriate for the modular formation of a university. The possibility that it might "be best to build separate institutes rather than to establish separate faculties all at once" had already been raised in the 1913 plan, and it stood at the center of the 1914 plan. But in both cases this was seen as a temporary measure on the way to a complete university. In 1920, however, the research institutes became a goal in and of themselves, with the ultimate ambition being to establish what we would today call a center of academic excellence.

There was no existing organizational model for the relationship between the secondary units and the overall institution in the proposed plan to begin with research institutes. In German universities, for example, the faculties were units of the larger institution, but they were principally teaching bodies. In research organizations, such as the

112 "Preparations Report" (above, note 111), p. 1.

Physikalisch-Technische Reichsanstalt (the National Metrology Institute) in Berlin, subsidiary institutes were meant to pursue the research program laid out by the larger organization. In Jerusalem, however, the research institutes would be only partly affiliated with the university, as their funding would come not from the university itself but from outside sources. Thus, the founders disregarded existing examples and fashioned a new organizational structure that, as the Preparations Report noted, was meant "to combine those elements in the continental system of State Universities with those in the English system of Independent Universities governed by Trustees as shall be found to fit in most naturally with conditions in Palestine."[113] According to Weizmann, the English system would produce a university functioning as a central administrative system for research institutes that would each be run independently by its own board.[114]

The ambition for scientific excellence was one shared by the founders of other universities as well. It was in the spirit of the academic policy mandating that the science departments be started "as Research Institutes and not as Teaching Faculties" that Weizmann declared at the opening ceremony in 1925 that the university's reputation must be built up "by the distinctive value of its contributions to the common stock of knowledge."[115] In fact, what was dedicated at that event was an

113 *Ibid.*, p. 5.

114 Weizmann to Wormser, *CWL*, X, no. 301 (December 5, 1921). See also Edward Shils and John Roberts, "The Diffusion of European Models Outside Europe," in Rüegg, *History of the University in Europe* (above, note 2), pp. 190–191.

115 See above, note 68. Here Weizmann shows his commitment to research for its own sake and to excellence in its pursuit. Others who also served as midwives for the new institution took the same position, among them Hayim Nahman Bialik, J.L. Magnes, and Einstein. For a discussion of this subject, see Shaul Katz, "The Scion and the Tree: The Hebrew University of Jerusalem and Its German Epistemological and Organizational Origins," in Marcel Herbst (ed.), *The Institution of Science and the Science of Institutions: The Legacy of Joseph Ben-David* (Boston Studies in the Philosophy and History of Science, 302), Dordrecht–Heidelberg–New York–London: Springer, 2014. Aryeh Goren notes that the opening ceremony in 1925 was in fact the fourth in a series of convocations on Mount Scopus, all of which were seen by the public as events marking the opening of the university.

institution that, despite being labeled a university, at least officially did not offer instruction; it was predicated on an ideal of excellence in scientific research and scholarship.[116] Thanks to Weizmann's central role in shaping the science policy of other research institutes in Israel, they were imbued with the same academic ideal, as we will see below.

On the Road to the Complete University, 1925–1935

The research institutes that comprised the Hebrew University at its inception were the Chemistry Institute, opened in 1924; two departments of medical research that were meant to become part of a faculty of medicine (the Department of Parasitology, opened in 1924, and the Department of Bacteriology and Hygiene, opened in 1926); and the Institute of Jewish Studies, whose inauguration, during Hanukah 1924, was seen as a ceremony marking the university's opening. In 1926 two more institutes were added, the School of Oriental Studies and the Institute of Palestine Natural History; the next year, 1927, saw the opening of the Einstein Institute of Mathematics.[117] During 1928, when "courses in philosophy, history, and literature" – an instructional framework that did not at first have the status of a research institute – began to be offered, Magnes and other leaders of the institution felt that it was now ready to begin an organized program of instruction. At the outset of the 1928–1929 academic year, the university's first faculty, the Faculty of

There had been an opening ceremony for the Institute of Jewish Studies during the Hanukah holiday in 5685 (December 1924), led by Magnes, and that had been preceded by Einstein's lecture on Mount Scopus in February 1923 and by the 1918 cornerstone-laying ceremony. See Arthur Aryeh Goren, "Judah L. Magnes and the Early Years of the University," in Katz and Heyd, *History of the Hebrew University*, I (above, note 69), pp. 363–385 (Hebrew).

116 Instruction nevertheless took place during these early years in some of the institutes, but it did not lead to a degree. See the diagram "Academic Expansion," in [Shlomo Zalman Schocken,] *The Hebrew University: Its History and Development* (second revised edition), Jerusalem: Hebrew University, 1942, p. 67 (Hebrew).

117 According to the above diagram and the chapter "History of the Hebrew University," *ibid.*, pp. 1–17.

Humanities, opened its doors. This involved a bit of compromise with the aspiration that all the university's institutes and departments would be headed by world-class scholars and scientists.[118]

In 1914, Weizmann had stressed the recruitment of young and promising scientists who would gain their reputations at the fledgling institution, because he estimated that only a handful of top scientists would come to Jerusalem – only those who would do so out of commitment to the Jewish people and a sense of connection to Eretz Israel.[119] During the 1920s a few major scientists and scholars joined the university, but most of them remained outside the country, serving as members of the Academic Council or the board of trustees rather than actively directing the research institutes in Jerusalem. Such was the case with Einstein[120] and his fellow physicist Ornstein. But Edmund Landau, one of the great mathematicians of his generation, came from the University of Göttingen to inaugurate the Einstein Institute of Mathematics on Mount Scopus, and before he returned to Göttingen half a year later, he recruited Abraham Halevi Fraenkel, one of the most important mathematicians in Germany at that time, as well as another prominent figure in the field, Michael Fekete.[121] After the imposition of the race laws in Italy in 1938, Weizmann was also able to bring in Giulio (Yoel) Racah, a leading theoretical physicist.[122] In addition to these few successes, Weizmann managed to recruit a number of young and promising scientists, among them the parasitologist Saul Adler in 1924,[123] the organic chemist Max

118 See Michael Heyd, "Between Nationalism and Universalism, Research and Teaching: The Early History of the Hebrew University," in Rivka Feldhay and Immanuel Etkes (eds.), *Education and History*, Jerusalem: Zalman Shazar Center, 1999, pp. 355–376 (Hebrew).

119 Weizmann to Magnes, *CWL*, VI, no. 350 (May 25, 1914).

120 On Einstein and the Hebrew University, see above, note 98.

121 Katz, "Berlin Roots" (above, note 6).

122 See Issachar Unna, "The Genesis of Physics at the Hebrew University," in Katz and Heyd, *History of the Hebrew University*, I (above, note 69), pp. 589–624; and see his article in this volume, "Wealth or Genius? The Struggle of Weizmann and Einstein for Academic Excellence at the Hebrew University."

123 Katz, "The Scion and the Tree" (above, note 115), pp. 121–123.

Frankel in 1926,[124] and Ladislaus (László) Farkas, a refugee from the University of Berlin who was considered a very gifted figure in the field of physical chemistry, recruited by Weizmann in 1934.[125]

Andor Fodor's appointment to head the Chemistry Institute was a

124 The recruitment of Frankel is mentioned in Weizmann's letter to Magnes, *CWL*, XIII, no. 94 (October 5, 1926). To ensure that Frankel and also the young chemist Nathaniel Thon could be retained, Weizmann informed Magnes that he himself would pay their salaries. Two young chemists who later had an important impact not only on local but also on world science, the brothers Aharon and Ephraim Katchalsky (later Katzir), took their first steps in the study of large molecules under Frankel. See, e.g., Ephraim Katzir, *Life Story*, Jerusalem: Carmel, 2008, pp. 55–62 (Hebrew). For more on Frankel, see Assaf Selzer, *The History of the Hebrew University of Jerusalem: Who's Who Prior to Statehood – Founders, Designers, Pioneers*, Jerusalem: Magnes, 2013, pp. 279–282 (Hebrew).

125 Michael Chayut, "From Berlin to Jerusalem: Ladislaus Farkas and the Founding of Physical Chemistry in Israel," *Historical Studies of Physical and Biological Sciences*, 24/2 (1994), pp. 237–263. The scientists cited here are just a few of those whom Weizmann sought to bring to the Hebrew University, and later – or already at the same time – to the Sieff Institute. Two scientific giants with whom Weizmann was in touch directly or indirectly were physiologist and biochemist Hans Krebs (1900–1981, Nobel laureate 1953) and physical chemist Michael Polanyi (1891–1976). For correspondence mentioning Krebs, see, e.g., Salman to Weizmann, CWA, 23-1736 (April 17, 1934). Regarding Polanyi, see Weizmann to Schmaltz, April 14, 1935, as well as Weizmann's correspondence with Polanyi, CWA, 14-1813. I am grateful to Eran Tal of the Weizmann Archive for locating this correspondence. Neither Krebs nor Polanyi accepted invitations to move to Jerusalem. The Hebrew University Archives preserve correspondence with Krebs, as well as a file of correspondence with Carl Neuberg (1877–1956), who was considered the founder of biochemistry. Weizmann pleaded with Neuberg to join the Hebrew University; he acquiesced after long negotiations but remained for only a short time. In the end he went to the Far East and from there to the United States (see the Neuberg file in the Hebrew University Archives). A respectably long entry on Neuberg can be found in the academic staff directory of the Hebrew University Yearbook for 1942. Neuberg was the only listed staff member who had served as the director of one of the Kaiser Wilhelm Research Institutes, in Berlin-Dahlem. During most of his time there he also served as director of that institution's Institute for Experimental Therapy. See *The Hebrew University: Its History and Development* (above, note 116), p. 138. For more on Weizmann's efforts to rehabilitate German Jewish refugees, such as his work with the Central British Fund for

different case. Fodor had served as *Privatdozent* and later achieved the rank of *Ausserordentlicher Professor* at the University of Halle. He accepted Weizmann's invitation to come to Mount Scopus and help found the Chemistry Institute. Weizmann originally viewed him as an ally and partner, and it is clear from his correspondence with Fodor and others regarding the Chemistry Institute that at the time, in the early 1920s, Weizmann intended to come to Jerusalem and play a role in the construction of the laboratories, to resume his research activities there, and to head the new institute. It is also obvious that Weizmann tied the establishment of the Chemistry Institute to that of the Microbiology Institute, thereby intimating the direction of the research he intended to pursue.[126] But neither of these things happened. In 1923 Fodor was made responsible for the construction and equipment of the Chemistry Institute, and in 1925 he became one of the first professors appointed by the Hebrew University. When he was named to head the Chemistry Institute, he was given the academic position that Weizmann had coveted since his days at the University of Manchester.

About a year after the opening ceremony in 1925, Weizmann was elected president of the Hebrew University and chairman of its Board of Governors. Einstein was elected chairman of the Academic Council, and Magnes was made chancellor, responsible for administration. This established three different sources of authority at the university's helm. The three of them sought to add a fourth officer, an academic chief or rector who would operate from Jerusalem, in the person of the recently arrived Edmund Landau; but Landau did not accept the post. A few

German Jewry, see Rose, *Chaim Weizmann* (above, note 56). [See also Shulamit Volkov's article in this volume. – Ed.]

126 Weizmann's first letter to Fodor is dated October 24, 1921; see *CWL*, X, no. 263. He responded warmly to the plans that Fodor had sent him (now kept in the Weizmann Archive) for the establishment of research institutes in Jerusalem, and he remarked that he had recently read one of Fodor's articles and would be happy to hear about the possibility of Fodor taking part in establishing the institutes. See also Reinharz, *CW*, II, p. 362; and Katz, "The Scion and the Tree" (above, note 115), pp. 126–130. In 1923, Fodor was appointed director of the Chemistry Department, certainly at his own insistence. See *Inauguration*, Large Format (above, note 67), pp. 10–11.

years later Weizmann proposed himself as academic chief (see below). These discussions came to an end in 1935, with the establishment of the university Senate. The Senate chose as rector one of its own members, Shmuel Hugo Bergmann – an elected academic director chosen by and acting on behalf of the university's professors. The appointed Academic Council was disbanded, thus terminating the position of its director, Albert Einstein. This completed the process of establishing a centralized autonomous academic administrative body.

Neither Weizmann, who lived in London, nor Einstein, who lived in Berlin (and did not even bother to come to Jerusalem for the inauguration ceremony), nor Magnes, who lived in Jerusalem, had experience in running a university. During its early years, the three of them agreed about the ideal of scientific excellence and believed that it should guide the new institution, but they could not agree on the practical steps needed to realize this goal, and it was not long before they were at loggerheads.[127] One prominent issue that divided them was systematic teaching. Magnes, the hands-on administrator, sought to introduce organized instruction. Weizmann, for his part, sought to preserve the research institute model; he imagined himself taking an active part in the research mission of the university he had established, while also playing a central role in its management, a position that would provide him

127 On the relations between Weizmann and Einstein and on Einstein's conflict with Magnes, see the works cited in note 98 above. [See also the articles by Hedva Ben-Israel and Issachar Unna in this volume. – Ed.] One of the results of these disputes was the establishment of the Survey Committee, also known as the Hartog Committee, headed by Sir Philip Hartog. "The Report of the Survey Committee of the Hebrew University of Jerusalem, 1934" called for far-reaching changes in the university's structure, which led, as noted above, to the establishment of the Senate and the election of a rector. Herbert Parzen was one of the first to investigate the establishment of the Hartog Committee and its recommendations; see idem, *The Hebrew University, 1925–1935*, New York: Ktav, 1974. Much on the Hartog Committee can be found in Katz and Heyd, *History of the Hebrew University*, I (above, note 69), the first of several volumes on the history of the Hebrew University. See also Uri Cohen, *The Mountain and the Hill: The Hebrew University in Jerusalem during the Pre-Independence Period and the Early Years of the State of Israel*, Tel Aviv: Am Oved, 2007, pp. 76–80 and elsewhere (Hebrew).

with academic and symbolic visibility. But during the 1920s Weizmann was fully involved in Zionist activity, and, to his chagrin, he had to set aside his scientific work and his role in shaping the university he had envisaged on Mount Scopus.[128] Weizmann had envisaged the Hebrew University as a project of national revival that would also provide an answer to his personal aspirations as a scientist; ultimately, however, he found himself without an institution at which he could satisfy his passion for science and fulfill his academic policy.

At the beginning of the 1930s, when it became apparent to him that his position in the Zionist Organization was weakening, he decided not to run for reelection to the movement's presidency. He saw this as an opportunity to resume his academic and research activity. Fifteen years had passed since he had left the University of Manchester, and since leaving his positions at the acetone factories, he had not maintained a regular connection with academic research. His desire to return to the world of science led him from Jerusalem to Rehovot.

III. From Jerusalem to Rehovot, from a Research and Teaching University to Research Institutes

How was it that Chaim Weizmann, the most enthusiastic promoter of the Hebrew University of Jerusalem and the chairman of its Board of Governors, left Jerusalem for Rehovot to found what might be seen as his own private scientific facility – the Daniel Sieff Research Institute, the nucleus from which the Weizmann Institute of Science grew –

128 Weizmann to Magnes, *CWL*, XIII, no. 94 (October 5, 1926). Although Weizmann offered Magnes considerable academic advice on the Chemistry Institute, as if he actually stood at its head, by then he clearly did not really intend to become part of it. Nevertheless, the claim that Weizmann did not engage in research during this period is not entirely accurate. In the years 1922–1925, Weizmann, together with his brother-in-law, Joseph Blumenfeld, registered ten patents that must certainly have come out of his research. [See nos. 35–44 in the list of Weizmann's patents and the article by Raphael Mechoulam in this volume. – Ed.]

which many, including Weizmann himself at times, viewed as a competitor and even rival to the university?[129]

To answer this question, we must investigate the following five issues: (1) The manner in which Weizmann's work with Paul Ehrlich in 1914 shaped his academic policy and, in particular, his academic ideals; (2) The lessons Weizmann the chemist and bacteriologist learned from the acetone project, and the implications of its success for his research and Zionist policies; (3) The undefined relationship between the Agricultural Experimental Station and the Hebrew University in its early years, especially regarding the role assigned to each in the establishment of what was to be a jointly managed agricultural college; (4) Weizmann's connections with the agronomist Yitzhak Elazari-Volcani (Wilkanski, 1880–1955), the head of the Agricultural Experimental Station, which eventually, with Volcani's blessing, became the nucleus of the Sieff Institute; and (5) Weizmann's longing to head a scientific academic institution and research laboratory in his field, so that he could realize his academic vision, pursue his scientific ideas, and find an outlet for his research energies.[130]

A number of Weizmann's biographers and some historians of the Sieff

129 The contention of some historians that the Weizmann Institute of Science was simply a continuation of the Sieff Institute is well validated by the important study of Sylvia Flowers, "The History and Development of the Weizmann Institute of Science," M.A. Thesis, University of London, 1975. The present account of the Sieff Institute's establishment and first years is also based on documents in the Chaim Weizmann Archive; see note 1. The account here ends with the last letter of Weizmann's that addresses matters relating to the Weizmann Institute, written to Dewey Stone; see *CWL*, XXIII, no. 355 (November 29, 1951).

130 "There were two factors which urged me on in this change [return to research]. First, my intrinsic relation to science, which had been part of my life since my boyhood; second, my feeling that in one way or another it had something to do with the building of Palestine. I was already thinking, then, of a research institute which should work in combination with the Agricultural Experiment Station at Rehovoth – and of something larger, and of wider scope too. And it was during the period when I was out of office that the Daniel Sieff Research Institute was founded, to be followed many years later by the Weizmann Institute of Science." *Trial and Error*, p. 422.

Institute have claimed that Weizmann turned to Rehovot because of an insult – the rejection, in 1931, of his proposal that he assume the position of academic chief of the Hebrew University while also directing his own laboratory on Mount Scopus.[131] But a more scholarly way of

131 Weizmann later wrote in his memoirs, "with some distaste," as he put it, about "my demission from the Presidency of the Zionist Organization at the Congress in July 1931" (*ibid.*, p. 417). Two months before his resignation, in his correspondence with Einstein about the idea of appointing mathematician Abraham Halevi Fraenkel as the Hebrew University's academic chief, Weizmann hesitantly broached the idea that he might be appointed to that post. It would have to be in tandem with his return to research, he asserted. "On July 1 [1931] I will be relieved of the obligations that so trouble me now [the presidency of the Zionist Organization], and I will want to devote myself to science once again. Throughout these past years I have made an effort – with great difficulty – to remain as current about developments as I could. As you know, my ideal has always been to come to Jerusalem .... From next month onward [I will be able] to remain in Jerusalem for most of the winter semester and to begin to work there .... We will be able to move to Jerusalem permanently when my son commences his studies at the university .... In this way I can begin some new projects in the chemistry laboratory and perhaps I could head the biochemistry department of our excellent experimental station (it has been offered to me!)." Weizmann to Einstein, *CWL*, XV, no. 152 (May 21, 1931). Compare Weizmann to Volcani, *CWL*, XIII, no. 90 (October 2, 1926). Several authors have cited this source and others like it, written by Weizmann and his wife, Vera, to claim that the founding of the Sieff Institute was Weizmann's reaction to the rejection of his suggestion to the Hebrew University. These writers did not, however, take into account Weizmann's multifaceted nature, offering himself as he did for ever more leading positions in the Zionist movement. Neither do they consider his intention of resuming personal involvement in laboratory research during this period. It should be noted that Weizmann, in the above-mentioned letter to Einstein, referred to a proposal that had already been made to him – that he head the biochemistry department of the Experimental Station. In other words, without any connection to his post in the Zionist Organization or his resignation from it, he was already thinking of a research project and laboratory in Rehovot. Without diminishing Weizmann's sense of affront, it seems difficult to argue that this was the only motivation behind the establishment of the Sieff Institute. For more on this period and this incident in Weizmann's life, and on the establishment of the Sieff Institute and its early days, see Rose, *Chaim Weizmann* (above, note 56), pp. 187–190.

looking at this development, one that puts less weight on Weizmann's emotions, brings to the fore other elements, both ongoing and transitory, relating to his efforts to bring his academic and research policies to fruition.

1. Weizmann and the model of the perfect research institute. Since the winter of 1914 and his collaboration with Ehrlich in drafting guidelines for the establishment of a "jüdisches medizinisch-naturwissenschaftliches Institut" in Jerusalem, Weizmann's goal had been to found a multidisciplinary research institute, what we would today call a center of scientific excellence. It seems likely that Ehrlich also provided Weizmann with a list of the organizational requirements for fostering an institute that could, within just a few years, take its place among the world's exemplary research institutions, one with the potential for producing a Nobel Prize winner. In line with this view, Weizmann thought it best that the university, at least in its first years, take the form of an umbrella under which several research institutes would operate. He did not achieve this aim. Nevertheless, his efforts to found such an institution continued, and his hope to head it did not diminish.[132]

2. Weizmann's vision of "technical bacteriology."[133] Weizmann's

132 In his speech at a ceremony marking the Hebrew University's tenth anniversary, Weizmann reconciled himself to making teaching part of the institution's mission. This was no doubt in response to the demand for university-level instruction on the part of "young Jews who crave learning and for whom the gates were closed" after the Nazi party's rise to power. See "Speech by Dr. Chaim Weizmann" in the article "Hebrew University's Ten-Year Anniversary," *Doar hayom*, April 11, 1935, p. 1 (Hebrew).

133 On the use of the term "technical bacteriology," see "The Daniel Sieff Research Institute in Rehovot," *Tekhnikah umada'*, year 2, no. 8(13) (September 1938), pp. 1–3 (Hebrew). The source for this article is the booklet *The Daniel Sieff Research Institute*, Rehovot: Palestine Publishing Company, 1938. There the term appears, for example, on p. 20, where the subjects of the institute's research projects are enumerated. The booklet seems to have been printed twice, first for the opening celebrations and then again four years later, with an added list of the articles published by the institute's staff during its first years. The Weizmann Institute Archive preserves a copy with the handwritten inscription "published autumn of 1938." About a decade later, referring to Weizmann's vision, Ernst David Bergmann used

sweeping claim that "Nature and bacteria will work for me" seems to have preceded the synthetic rubber project that began in 1910, as related in Chapter I above. That project, aimed at using bacterial fermentation products, in principle represented the success of Weizmann's "biotechnological" idea, and the acetone project even more so.[134] But Weizmann did not exploit these successes as a springboard for new research programs or projects, or even for a bacteriological (or what we would today call biotechnological) vision. Presumably, the reason was Weizmann's growing dedication to and preoccupation with Zionist affairs. Nevertheless, it is reasonable to suppose that, with the success of the acetone project and, certainly, the widespread adoption in the 1920s of the innovative concept of bacterial fermentation products as a source of chemicals sought by other industries, new directions of investigation inspired him. These ideas would form the basis of the biotechnological and biotechnological-political vision he later articulated.[135] But to

the then-current term "chemurgy," referring to the branch of chemistry concerned with the industrial use of organic raw materials, especially of agricultural origin; see Ernst D. Bergmann, "Dr. Chaim Weizmann and the Weizmann Institute of Science," *Yedi'ot kalkaliyot*, 2 (February 1950), pp. 80–82 (Hebrew). Robert Bud discusses the development of the chemurgy movement in the 1930s; see Bud, *Uses of Life* (above, note 30), pp. 46–50. He does not use the term "technical bacteriology."

134 On this see note 191 below.

135 It may well be that evidence of such research plans, and perhaps also a formulation of such a vision, lies in the "technical material" he produced, which has not yet been published; see Weisgal, "Introduction," *CWL*, VI. A letter Weizmann wrote to Abraham Halevi Fraenkel at the end of 1932 shows that he was involved in, or intended to pursue, a bacteriological research project for which he required "yeast, in clean cultures," as well as other cultures. He asked Fraenkel, who was in Kiel at the time, to obtain the needed cultures for him from a research institute in that city that specialized in fermentation: "I am attempting to establish a bacteriological museum for the Experimental Station … excuse me for putting you to so much trouble, but it is for Eretz Israel. I am busy with an agricultural college of which I will in practice become the director, God willing." Weizmann to Fraenkel, *CWL*, XV, no. 322 (December 5, 1932).

realize this, fully or partially, Weizmann needed a large laboratory – in fact, a research institute of his own.[136]

3. The plan for the Hebrew University's agricultural college. The Institute of Agriculture and Natural History–Agricultural Experimental Station (see Chapter I of this article), the first research institute established in Palestine by the Zionist Organization,[137] opened its doors in Tel Aviv in 1921–1922 under the directorship of Otto Warburg (1859–1938), with Yitzhak Elazari-Volcani heading the Agricultural Experimental Station. In its five years of activity, the institute's professional staff gradually expanded to about forty.[138] Weizmann had first explicitly broached the idea of an agricultural studies institute in 1918, as part of the complete university proposal published in *On the Threshold.* When the university opened in 1925, the Institute of Agriculture and Natural History became one of its components, along with the Institute of Jew-

136 Weizmann registered a number of patents in the 1920s; see above, note 128. Beginning in the early 1930s, Weizmann ran a private laboratory in London: "I opened up a modest laboratory at 6 Featherstone Buildings, Holborn .... The laboratory was not particularly well-fitted out, but it served my purposes, at least at the beginning. I also linked up again with an old friend and assistant, Mr. H. Davies, who had been with me in Manchester and who had worked with me during the first years of World War I. It began to look again like old times. I enjoyed immensely going to the laboratory every day and returning home in the evening. It reminded me very much of my years at Manchester University." *Trial and Error*, p. 422. In August 1940, Weizmann submitted to his accountants a list of the research subjects addressed in the laboratory, the names of the workers, and a detailed account of the operating costs; Weizmann to J. Cameron, *CWL*, XX, no. 30 (August 22, 1940), pp. 33–34. Another laboratory that was available to Weizmann from the beginning of the 1940s is mentioned in note 191 below.

137 Shaul Katz, "Sociological Aspects of the Growth and Turnover of Agricultural Knowledge in Israel, 1880–1940," Ph.D. Dissertation, The Hebrew University of Jerusalem, 1986, pp. 248–262 (Hebrew).

138 "The Zionist Organization, Institute of Agriculture and Natural History, Agricultural Experiment Station, First Report Covering a Period of Five Years, 1922–1926, Tel Aviv, Palestine, November 1926" (the Warburg–Volcani Report), p. v, "Organization." Known in the course of its history as the "Agricultural Experiment Station," the "Experimental Station," and the "Agricultural Experimental Station," it eventually became the Agricultural Research Organization – Volcani Center.

ish Studies, the Chemical Institute, and the Microbiological Institute.[139] But in 1925, making the Institute of Agriculture organizationally a part of the university was still an aspiration. It took fifteen years for the Institute for Agricultural Studies to get off the ground as a joint operational framework of the Agricultural Experimental Station and the University.[140] Weizmann and Volcani were among the leading figures involved in establishing this joint framework, at least in its early stages.

4. The Weizmann-Volcani connection. From the time he arrived in Palestine at the head of the Zionist Commission in the spring of 1918, Weizmann maintained close contact with Yitzhak Elazari-Volcani, the agronomist who was, certainly after the death of Aaron Aaronson, one of the leading local experts on settlement and agriculture.[141] Weizmann valued Volcani's professional opinions on these subjects. As the person responsible for all research and higher education in the Yishuv – the organized Jewish community in Palestine – Weizmann looked out for the interests of the Agricultural Experimental Station as well.[142] It is also

139 *Inauguration*, Large Format (above, note 67), pp. 9–13. Under the heading "Relation to the University," the text reads, "It is now intended to attach the Institute to the University as the basis of a University College of Agriculture." The Hebrew version (p. 11) refers only to an "agricultural institute." The Arabic text was not examined. For *On the Threshold* see above, note 102.

140 "Historical Review," in *The Hebrew University of Jerusalem, Faculty of Agriculture*, Rehovot, 1958, pp. 13–40 (Hebrew); Moshe J. Pinthus, "The Study of Agriculture: Vision to Reality," in Hagit Lavsky (ed.), *The History of the Hebrew University of Jerusalem*, III: *Academic Progression in a Period of National Struggle*, Jerusalem: Magnes, 2009, pp. 217–251 (Hebrew). The Institute of Palestine Natural History was established when the biology units of the Agricultural Experimental Station and most of its researchers, led by Otto Warburg, moved from Tel Aviv to Jerusalem beginning in 1926.

141 *Festschrift in Honor of Professor Elazari-Volcani*, Rehovot: Agricultural Research Station, 1950 (Hebrew); Natan Bistrzycki (ed.), *The Guide: In Honor of Yitzhak Volcani's Seventieth Birthday*, Jerusalem: Zionist Executive–Jewish National Fund, 1950; Y. Reichart, "Volcani's Enterprise," in *Hebrew University of Jerusalem, Faculty of Agriculture* (above, note 140), pp. 263–265 (Hebrew).

142 The Warburg–Volcani Report (above, note 138) was submitted to Weizmann, who was president of the Zionist Organization. An example of Weizmann's special concern for the Experimental Station can be found in his report to Volcani

worth noting that, during the months that Weizmann spent in Palestine, he and his associates in the Zionist movement established a political alliance with the Hapo'el Hatza'ir party, of which Volcani was a leading spokesman. This alliance was based on their common belief in political moderation, which both had held in the past and would continue to hold for many years to follow.[143]

5. Weizmann's aspiration to head a research institute in the Yishuv. In April 1925, at the first meeting of the committee that was to become the Hebrew University Board of Governors, the issue of an agricultural college came up for discussion. It was decided to postpone the project until the Agricultural Experimental Station moved from Tel Aviv to its permanent home in Rehovot.[144] But at the Board of Governors meeting

on the possibility of receiving funding for its development from the Imperial Research Committee: Weizmann to Wilkanski-Volcani, *CWL*, XIII, no. 90 (October 2, 1926). Another example is their discussion of Volcani's November 1931 plan to found a research institute in the framework of the Experimental Station in Rehovot, which would be named for Weizmann and devoted to the study of natural products – an idea that he passed over to his advisers, David Eder and Leonard Stein; see Weizmann to Wilkanski-Volcani, *CWL*, XV, no. 251 (February 2, 1932). Cf. Weizmann's letter to Einstein, *CWL*, XV, no. 152 (May 21, 1931), in which he declared that he was no longer interested in maintaining ties with the university but doubted that it would be possible to establish a large research institute in Rehovot.

143 Meir Chazan, *Moderation: The Moderate View in Hapo'el Hatza'ir and Mapai, 1905–1945*, Tel Aviv: Am Oved–Chaim Weizmann Institute for the Study of Zionism and Israel, Tel Aviv University, 2009 (Hebrew).

144 According to the inaugural booklet, the Institute was to be attached to the university as an agricultural college; *Inauguration*, Large Format (above, note 67), p. 14. At the Board of Governors meeting, however, Warburg advised Magnes to postpone establishment of an agricultural college until the Experimental Station, which was to be located in Rehovot, could ensure the necessary support. This exchange appears in the collected minutes of the Board of Governors' meetings in the Hebrew University Archives. There seem to be other versions of the minutes; for example, see note 146 below regarding the fifth meeting in August 1929, whose minutes are not in the Hebrew University Archives. In referring to that meeting, the authors of the "Historical Review" (above, note 140) seem to have used original documents and to have been faithful to the source.

in 1929, even before the station completed the move in 1932,[145] the question was raised again. In August 1932, the Board of Governors resolved to establish a College of Agriculture and to appoint Weizmann to head it. But there is reason to believe that the idea of appointing Weizmann to head the college, tying the appointment to the provision of a laboratory for him, and expecting him to raise the necessary funds, was first floated in 1929, if not earlier.[146] Whatever the case, Weizmann once again began planning the establishment of the new institution by studying information on the desired academic content and seeking sources of financial support.[147] In his search for donors, Weizmann

145 "Station Organization and Location of Activities," *Third Report on Five Years of Activities*, Rehovot, 1925, p. 14 (Hebrew). I am grateful to Amalia Barzilay of the Agricultural Research Organization (ARO)–Volcani Institute for providing me with this document and others touching on the station and on the Sieff Institute. The biological research staff headed by Warburg was incorporated into the Hebrew University's Institute of Palestine Natural History, founded in 1926.

146 According to the "Historical Review" (above, note 140), "At the fifth meeting of the Board of Governors, which convened in August 1929, it was decided to authorize the University Council to study the question of the relations between the University and the Agricultural Experimental Station, with the intention of establishing an Institute of Agricultural Studies. In the wake of this decision, informal talks were conducted in 1929 between representatives of the University and the Station" (p. 18). At the Board of Governors meeting in August 1932, "the Board decided that it was time to commence with the establishment of the Institute of Agricultural Studies .... The school will be headed by C. Weizmann, who will – so it is hoped – take up his post in October 1933. Special funds will be raised to establish the Institute .... Additional funds will be raised in the sum of P£2,000 (Palestine Pounds) per year for the construction and outfitting of new laboratories, as well as P£1,000 per year, for four years, to install a chemistry laboratory for C. Weizmann" (*ibid.*, pp. 19–20). So, too, under the heading "Agricultural College" in the "Decisions of the Academic Council and Board of Governors of the Hebrew University at the Meeting Held in London on August 14–15, 1932," p. 2. And see also Pinthus, "Study of Agriculture" (above, note 140).

147 For example, Weizmann met with Sir John Russell, an expert on soil science and the director of the Rothamstead Station, the first agricultural research facility in England. Russell had toured Palestine a few years earlier and written on its agriculture. Weizmann also consulted with Sir Phillip Hartog, whom he would later recruit to head a committee to study and make recommendations about the

learned that the Sieff family and the allied Marks and Sacher families, all of whom he knew from his Manchester days, were prepared to fund the establishment of a research institute in memory of the Sieff family's young son, Daniel. It would be called the Daniel Sieff Research Institute, and Weizmann would head it.[148]

At the beginning of 1933, and perhaps even earlier, Weizmann initiated the process of constructing the buildings and laboratories for the new Institute, which was to be located within the Agricultural Experimental Station compound then being constructed in Rehovot, as an integral part of it. Volcani was active as one of the new institute's major facilitators and as a promoter of Weizmann and his initiatives.[149] Weizmann added the director of the station's chemistry department, Isaac Magasanik, to

Hebrew University's governance; see above, note 127. In a speech addressed to the University community on December 22, 1932, Magnes referred to "the Institute of Agriculture," whose program the university and the Experimental Station were planning for the next year. Magnes did not mention the post intended for Weizmann. See *Speeches of the Chancellor of the Hebrew University*, Jerusalem: Hebrew University, 1936, p. 144 (Hebrew).

148 The name of the institute was engraved in Hebrew, English, and Arabic over its main entrance and appeared in all three languages on its letterhead: מכון למחקר ע"ש דניאל זיף – The Daniel Sieff Research Institute – معهد أبحاث دانيال سيف. In Hebrew the name was originally written זיף rather than זיו, as is now standard, but the latter spelling appears in the institute's 1941 Memorandum of Association: The Companies Ordinance, Memorandum of Association of the Daniel Sieff Research Institute, May 1, 1941, CWA, Sieff Institute Container. For the story of how the idea of commemorating Daniel Sieff by means of a research institute came about, see Israel Sieff, *Memoirs*, London: Weidenfeld and Nicolson, 1970, pp. 131–133.

149 See Volcani's address at the Sieff Institute's opening ceremony, in the *Daniel Sieff Research Institute* booklet (above, note 133), pp. 17–19. See also Yitzhak Elazari-Volcani (A. Tziyoni), "With the Spirit Level" (Nissan 5697 [1927]), in idem, *Collected Writings*, V: *Aspects*, Tel Aviv: Tversky, 1954, pp. 148–161 (Hebrew). This article, a paean to Weizmann, analyzes his leadership and surveys how it was expressed. Volcani seems to have been the first person to consider the connection between Weizmann's Zionism and his work as a scientist: Weizmann, he wrote, "transferred the values of his chemistry laboratory to the laboratory of life: an acknowledgment of organic laws, the art of refining, the theory and practical use of synthesis, and above all, a sense of proportion." *Ibid.*, p. 148.

the emerging Institute's founding team, further evidence that Volcani and his circle believed that Weizmann's new institute would be part of the station.[150] Weizmann was finally able to establish the kind of research institute he desired.

While planning the institute's construction in Rehovot, Weizmann made inquiries about candidates for research positions. Because of the increasingly ominous tenor of events threatening Germany's Jews, Weizmann had no great difficulty recruiting refugee researchers who needed a scientific haven. However, his success on this score should be placed within the larger framework of his efforts to augment the Yishuv's intellectual strength and technical prowess. He was successful in attracting promising junior scientists to the Sieff Institute but was deeply disappointed in his efforts to enlist more renowned scientists for either the Hebrew University or the fledgling institute. Among those he attempted but failed to convince were Albert Einstein, James Franck (1882–1964, Nobel laureate 1925), Richard Willstätter, and Georg Placzek (1905–1955),[151] who declined his invitation to join in the Zionist enterprise "for our benefit as well as theirs."[152] But some of these

150 On Magasanik, see Aharon Bondi, "The Late Dr. Isaac Magasanik," *Hasadeh*, 21 (1940/41), p. 353 (Hebrew). I am grateful to Amalia Barzilay for this reference.

151 Stuart A. Rice and Joshua Jortner, *James Franck, 1882–1964: A Biographical Memoir*, Washington: National Academy of Science, 2010, http://jfi.uchicago.edu/aboutinstitute/documents/jfranck-memoir.pdf (accessed December 7, 2016). On Placzek, see Unna, "Genesis of Physics" (above, note 122), and Chayut, "From Berlin to Jerusalem" (above, note 125).

152 See *Trial and Error*, pp. 417–426. But given Weizmann's success in winning over Ladislaus Farkas and Giulio Racah, and in bringing Carl Neuberg to Palestine for a short period (see above, note 125), it is clear that not all his efforts were for naught. This is all the more true if one counts Fritz Haber's intention of coming to Rehovot [on which see Shulamit Volkov's article in this volume. – Ed.]. In any case, the brief list noted here certainly does not exhaust the array of scientists Weizmann tried to interest. The persecution of Jewish scientists in Germany is discussed by Ute Deichmann, "The Expulsion of Jewish Chemists and Biochemists from Academia in Nazi Germany," *Perspectives on Science*, 7/1 (1999), pp. 1–86, https://muse.jhu.edu/journals/perspectives_on_science/v007/7.1deichmann.html#tab03 (accessed December 7, 2016); and by Yoav Gelber, *New Homeland:*

major figures were nevertheless willing to help Weizmann establish the Sieff Institute, especially by recruiting scientists to join it. Weizmann's first recruit, for the position of scientific director, was Ernst David Bergmann (1903–1975).[153] Weizmann's enlistment of Bergmann had major implications far beyond the management of the Sieff Institute and the

The Immigration and Absorption of Central Europe's Jews, 1933–1948, Jerusalem: Yad Izhak Ben-Zvi–Leo Baeck Institute, 1990, pp. 454–462 (Hebrew).

153 Weizmann's first letter to Bergmann is dated May 7, 1933: *CWL*, XI, no. 370. Carl Neuberg helped draw Bergmann to Weizmann's attention; see Weizmann to Neuberg, *CWL*, XV, no. 378 (May 9, 1933). Neuberg originally recommended that Bergmann be appointed to the Department of Chemotherapy at the Hebrew University (at Hadassah Hospital). This correspondence is a good example of how Weizmann solicited and obtained recommendations of junior scientists from senior figures with whom he had previous acquaintance. Regarding Bergmann, Weizmann also solicited the recommendations of Willstätter and Wilhelm Schlenk (1879–1943), Emil Fischer's successor at the University of Berlin and head of the Institute of Chemistry there. See Thomas T. Tidwell, "Wilhelm Schlenk: The Man behind the Flask," *Angewandte Chemie*, International Edition, 40/2 (Jan. 19, 2001), pp. 331–337, http://onlinelibrary.wiley.com/doi/10.1002/1521-3773(20010119)40:2%3C331::AID-ANIE331%3E3.0.CO;2-E/full (accessed December 7, 2016). Bergmann's first steps in his academic career as an organic chemist were taken under Schlenk; he and Herman Mark (see below) were two of Schlenk's Ph.D. students. Weizmann reported to Einstein on the recruitment of Bergmann and on the recommendations he had received; see Weizmann to Einstein, *CWL*, XV, no. 409 (June 8, 1933). Weizmann took this opportunity to urge Einstein to help recruit chemists and physicists for the university. On the first exchange of letters between Bergmann and Weizmann and its context, see William B. Jensen, Henry Fenichel, and Milton Orchin, *Scientist in the Service of Israel: The Life and Times of Ernst David Bergmann (1903–1975)*, Jerusalem: Magnes, 2011, pp. 34–43. In the short CV dated February 22, 1952, submitted by Bergmann to the Hebrew University, he notes: "At the time of my dismissal [from the University of Berlin], I was the first candidate for a professorship in organic chemistry at the Technische Universität Berlin at Charlottenburg," where Weizmann, it should be recalled, began his own career in organic chemistry. The two-page Hebrew document was typewritten on stationery of the Ministry of Defense, Division of Research and Planning; it can be found in the Hebrew University Archives, Bergmann file.

establishment of the Weizmann Institute; it would ultimately affect the entire research system in Israel.

The Sieff Institute – Establishment, Character, Research Program

The Sieff Institute was a research organization different from any other in the Yishuv of the 1930s. Unlike the Hebrew University and the Agricultural Experimental Station, it was neither founded nor funded by the Zionist Organization, nor was it meant to provide services to the public, committed as it was to basic rather than applied research. While formally a component of the Hebrew University's agricultural college, it was in fact a privately funded and administered scientific institute.[154]

The Sieff Institute was the realization of Weizmann's dream of a research institute, headed by himself, that would combine scientific excellence based on superior basic research with ramifications for agriculture and industry, and an emphasis on biotechnology.[155] The Sieff Institute was meant to accommodate the laboratory work Weizmann would carry out during his stays in the country and the projects in

154 The Institute for the Study of Hebrew Poetry, founded by Salman Schocken, was a private institute of a similar sort, with a like connection to the Hebrew University. Schocken was a leading figure on the university's board, see Anthony David, *The Patron: A Life of Salman Schocken, 1877–1959*, New York: Metropolitan Books, 1963.
With the rise of Nazism in Germany, Schocken and his agents were able to obtain the libraries and books of individuals and communities and bring them to Jerusalem. Ita Heinze-Greenberg cites testimony from the memoirs of Louise Mendelsohn, wife of architect Erich Mendelsohn, according to which it was Weizmann's invitation to her husband to design the Weizmann House in Rehovot that prompted Schocken to ask him to plan what would become the Schocken Institute for Jewish Research in Jerusalem and his adjacent private home, the Schocken villa. See Ita Heinze-Greenberg, "The Impossible Takes Longer: Planning the Weizmann House in Rehovot," *Cathedra*, 72 (June 1994), pp. 94–112 (Hebrew).

155 "Scheme for Daniel Sieff Research Institute" (above, note 100). Weizmann opens this memorandum by declaring that he intends to spend at least four months annually in Palestine in the years to come. The editor of the papers notes that Weizmann did not in fact realize this intention; *ibid.*, p. 33, note 1.

which he was interested, which would continue during his periods of absence. He would be helped by his loyal laboratory assistant from his time working on the synthetic rubber project in Manchester, Harold Davies (1886–1959),[156] and by Bergmann who, alongside his post as the institute's first scientific director, became Weizmann's confidant, ally, and scientific collaborator – almost a son.[157]

Weizmann viewed crops as a natural resource no less than coal and oil; thus, as in his acetone project, he turned to fermentation techniques on plants to produce feedstock for industry and for consumer goods in short supply. The basic concept also prompted Weizmann to envision the biological production of fuel, a project with enormous political and international implications, given the West's growing dependence on oil. On top of this, he believed that malnutrition in Asia and Africa could be ended by producing edible artificial proteins or their components.[158] Bergmann eagerly adopted and elucidated the principles of Weizmann's academic policy (emphasizing scientific research as a goal in and of itself) and his research policy (emphasizing fermentation projects and the vision of biological fuel), including their political ramifications.[159]

156 See above, note 43.

157 Weizmann used to call Bergmann "my only son"; see Meyer Weisgal, *So Far: An Autobiography*, New York: Transaction, 1971, p. 274. Weizmann recounts how he recruited Bergmann for the Weizmann Institute in *Trial and Error*, pp. 440–441.

158 Bergmann noted that the idea of using yeast enzymes to break down proteins could be found in the work of Willstätter and his students; see Ernst Bergmann, "Scientific Research and Social Construction," in Meyer W. Weisgal (ed.), *Chaim Weizmann: Statesman, Scientist, Builder of the Jewish Commonwealth*, New York: Dial Press, 1944, pp. 257–264. Given the absence of direct testimony from Weizmann himself about how he arrived at his ideas, Bergmann's remark contains a hint for further study that touches on Weizmann's relations with Willstätter, and Weizmann himself hints at the same; see *Trial and Error*, pp. 421–422. [See also Section IV in the chapter "Unpublished Passages" in this volume. – Ed.]

159 Bergmann, "Scientific Research and Social Construction" (above, note 158). A rather dramatic expression of Bergmann's identification with Weizmann and his research and academic policies can be found in idem, "The Daniel Sieff Research Institute in Rehovoth" (January 16, 1944), four typewritten pages, WIA, Bergmann file, 6/74, 94. This seems to have been intended as an article marking the

In keeping with the medical research institute model of 1914, the new institute was to be staffed with researchers of proven abilities and headed by a scientist of outstanding academic stature. Weizmann must certainly have viewed himself as a fitting candidate. While he had not for some time held a post at an academic institution or even been a regular member of a specific academic community, he had kept up his connections with contemporaries who belonged to the elite in his field, such as Robert Robinson. He had also nurtured new affiliations with such figures as physical chemist Jacques Errera (1897–1977), a participant in the Solvay Conferences, Fritz Haber (1868–1934, Nobel laureate 1918), and Richard Willstätter.[160] Weizmann solicited both their own participation and their recommendations for top scientists to staff the institute.[161] But, as already noted above, Weizmann found it difficult to persuade senior scientists and Nobel Prize laureates to join the Hebrew University or the new institute. Willstätter, who harbored deep respect for Weizmann and his work, came to Rehovot for the Sieff Institute's opening ceremony but declined the invitation to join the institute or head it. Haber, who eventually accepted Weizmann's offer, died in Switzerland before he could make the move.[162]

Sieff Institute's tenth anniversary, but it was never published. Bergmann expressed similar sentiments years later in his article "Science in Israel Twenty Years after Independence," *Mada'*, 13 (summer 1968), pp. 74–77 (Hebrew).

160 Weizmann approached Willstätter in 1932 at the Royal Society event in London at which Willstätter received the Davy Medal "for his distinguished researches in organic chemistry": https://royalsociety.org/grants-schemes-awards/awards/davy-medal/ (accessed January 31, 2017); see also *Trial and Error*, pp. 421–422. Robert Robinson and Arthur Lapworth were awarded the medal in 1930 and 1931, respectively; on them see above, at notes 16–18, and the notes there. Weizmann valued his connection with Willstätter and mentioned their personal friendship and plans for joint research on several occasions. See, for example, *Trial and Error*, pp. 421–422 [and see Shulamit Volkov's article in this volume. – Ed.].

161 In the "Scheme for Daniel Sieff Research Institute" (above, note 100), Weizmann listed the names of these scientists, their achievements to date, and who had recommended them.

162 On Haber's intention to accept Weizmann's invitation to come to Rehovot, see *Trial and Error*, pp. 434–438, and elsewhere. See also Dietrich Stoltzenberg, *Fritz*

In planning his research institute, Weizmann sought to integrate two approaches to research. In the first, standard in academic research organizations, scientists were expected to work according to their discipline's internal agenda – that is, to conduct scientific research for its own sake (pure research, or what we today call basic research) – and to publish their findings in scientific journals, under their own names and that of their institution. In the second, the accepted approach in goal-oriented research organizations, scientists took their agendas from outside the discipline. Recall that Weizmann spent most of his years as a working scientist at universities that integrated basic research with research aimed at the needs of industry.

From the Sieff Institute's inception, Weizmann laid out three spheres of activity for its staff: (1) agriculture-related problems; (2) industry-related problems, especially pharmaceuticals and fermentation; and (3) questions of a purely theoretical and scientific nature. The Institute, he declared, must stress the third sphere, because it first had to earn a reputation as a producer of distinguished scientific work.[163] Weizmann spoke in this vein at the institute's opening ceremony on April 3, 1934.[164]

Haber: Chemist, Nobel Laureate, German, Jew, Philadelphia: Chemical Heritage Press, 2004, Chap. 13. [See also the discussion of this issue in Shulamit Volkov's article in this volume. – Ed.] Haber's family bequeathed his library to the Sieff Institute, where it became one of the new institute's most important resources for researchers.

163 See "Scheme for Daniel Sieff Research Institute" (above, note 100), p. 4. Bergmann understood Weizmann's position to be that pure research would eventually produce useful products. He wrote that Willstätter's address at the Sieff Institute's inauguration stressed the idea that the approach to all problems, including practical ones, needs to be pure research, a standard that Weizmann never tired of emphasizing. See Bergmann, "Scientific Research and Social Construction" (above, note 158).

164 "Opening of Daniel Sieff Research Institute," Speech at Public Ceremony, Rehovot, April 3, 1934, *CWP*, II, pp. 41–44. In this speech, Weizmann stressed his long friendship with Volcani and expressed his gratitude for the latter's assistance in establishing the Sieff Institute as a component of the Agricultural Experimental Station. Other speakers at the event also underscored the importance of pure research, as did the philanthropist Israel Sieff in the letter he sent to the ceremony.

But, alongside these general statements, Weizmann's plan also gave special attention to the innovative pure research field of photochemistry, of which he spoke in a manner reminiscent of his earlier statement on his biotechnological work: "Let nature and bacteria work for me and produce synthetic rubber!" In photochemistry, Weizmann declared, "the principal reagent is the sun," of which there was an abundance in Palestine.[165] Was Weizmann, in 1934, proclaiming a vision of solar power, which became one of the scientific aspirations of the young State of Israel? Whatever the case, Yehuda Hirshberg, a physical chemist who had served as a scientific assistant to Jacques Errera, would pursue this subject.[166] More importantly, Weizmann's evocation of photochemistry as a pure research field is an example of his view that research for its own sake would ultimately be of practical use."[167]

Construction on the Sieff Institute began in the spring of 1933, and the abundantly equipped modern facility opened officially in April 1934. As at the Hebrew University's opening ceremony nine years earlier, the guests included leading figures in the British administration and the Zionist movement, as well as leading scientists from around the world.[168]

See *Haaretz*, April 4, 1934, which published several articles on the opening. See also "The Daniel Sieff Research Institute," *Tekhnikah umada'* 2/8 (above, note 133).

165 For the quote "Let nature and bacteria work for me and produce synthetic rubber," see above, note 37. For the expression "the principal reagent," see "Scheme for Daniel Sieff Research Institute" (above, note 100), p. 37, remarking on the position intended for Yehuda Hirshberg.

166 "Scheme for Daniel Sieff Research Institute" (above, note 100), p. 37. Presumably, as Weizmann turned most of his laboratory work in the direction of bacteriology, he began to see physical chemistry, and thus its practitioners, as a very important scientific research resource.

167 *Ibid.*

168 *Haaretz*, April 4, 1934 (Hebrew). Willstätter was the guest of honor and officially inaugurated the institute, taking the role Lord Balfour had played a decade earlier for the Hebrew University. Willstätter's great admiration for Weizmann was evident in the references he made to the latter in his keynote address, which took the form of a lecture on the history of science and the benefits that pure research had produced for humankind. Willstätter called on the institute's staff to focus on

There is no better indication of Weizmann's intention of making himself a permanent inhabitant of Rehovot than the construction of his villa near the Sieff Institute.[169] Weizmann did not find it difficult to give his enterprise in Rehovot historic symbolism. Rehovot, he pointed out, was not far from historical Yavneh, which became the seat of the Sanhedrin, the assembly of Jewish sages, after the destruction of the Second Temple.[170] Weizmann's original goal had been to get himself a laboratory in Rehovot; in the end he gained a research institute.

Weizmann began working there along with thirteen other young scientists, men and women, nine of whom were refugees from Germany.

in-depth research and to eschew publishing results that largely repeated what had been published previously. He referred to Weizmann as his great friend and confessed that he would be willing to exchange 100 of the articles he himself had written for one of Weizmann's, the name of which he never noted. Cf. Abraham Halevi Fraenkel, "Dr. Chaim Weizmann and the Development of Natural Sciences at the Hebrew University of Jerusalem," *Yedi'iot ha'Universitah ha'Ivrit biYrushalayim*, 2/3 (Dec. 1952), pp. 5–6 (Hebrew). Fraenkel's remarks recall Willstätter's: "While Weizmann has not been able to publish much in a quantitative sense, busy as he has been with the affairs of his nation and land, among his articles there is one that in my opinion is of equal weight to my entire life's work." [See Willstätter's letters to Weizmann in the appendix to Shulamit Volkov's article in this volume. – Ed.]

169 On building Weizmann's villa, see Heinze-Greenberg, "The Impossible Takes Longer" (above, note 154). The Sieff Institute building was designed by Benjamin Chaikin. In the years prior to World War II, the institute's annual budget was in the area of £12,000–£14,000; see Weizmann to Sieff, *CWL*, XII, no. 155 (April 24, 1946), p. 131.

170 The motif of the geographical proximity between historical Yavneh and the Sieff Institute (and later the Weizmann Institute) was echoed at various opportunities by the institutes' leaders. It was clearly a way of endowing the choice of Rehovot as the site of these centers of learning with national-historical significance. See, e.g., Weizmann, "The Role of Science," Address to the American Committee for the Weizmann Institute, New York, November 27, 1945, *CWP*, II, no. 83. The use of the phrase "Give me Yavneh and its sages" as the epigraph for *Eine jüdische Hochschule*, the early (1902) proposal for a Jewish institution of higher education (see above, note 69), expresses a national-intellectual aspiration. Of course, as the present chapter shows, there were other reasons why the Sieff Institute was located in Rehovot.

But, having achieved his dream of a research institute under his direction where he and colleagues would pursue their research, he did not end up spending much time there. In 1935 he once more took up the reins of the Zionist Organization and spent much time traveling to promote the Zionist cause and raise funds. Bergmann oversaw the scientific work at the institute, while Weizmann handed administration over to an appointed trust, most of whose members were not on the institute's staff,[171] and later to an executive committee made up of staff members.[172]

171 Weizmann proposed that the trust be composed of two representatives appointed by the donor family – Sir John Russell and Arthur Ruppin – with Louis Green as treasurer and Shalom Horowitz as legal counsel. "It would be in the interests of harmonious cooperation with the Agricultural Experimental Station," he wrote, "if Dr. Wilkansky [Volcani, the station's head] were invited to serve on the Trust" ("Scheme for Daniel Sieff Research Institute" [above, note 100], pp. 37–38). Later, certainly by the time the Weizmann Institute was established, the areas of the Agricultural Station that were allocated to the Institute had expanded. It was only in 1948, years after the station was largely reconstituted as the Hebrew University Faculty of Agriculture, that the compound was officially transferred to the Weizmann Institute.

172 Each committee member was assigned responsibility for a given area – research affairs, publications and library, finances and secretariat, instruments and upkeep. Isaac Magasanik, of the Agricultural Experimental Station, was appointed to oversee cooperation between the two organizations. Nevertheless, Weizmann asserted, "overall supervision of the scientific work of the entire Institute remains in my hands," letter in the form of a memorandum, May 8, 1936, WIA, Sieff Institute files, 1936–1940, 5/74, file 221. In this context, it is interesting to note Bergmann's description of the institute's staff: "A group of young scientists, under the leadership and inspiration of Dr. Chaim Weizmann, have made themselves into a kind of 'scientific *kvutzah*' [kibbutz] and devote their lives, knowledge, and abilities to the mission assigned to them" (Bergmann, "The Daniel Sieff Research Institute," above, note 159). There is, however, considerable evidence attesting to tension at the institute; see, e.g., Bergmann to Dr. Benjamin M. Bloch, December 27, 1936, WIA, Sieff Institute files, correspondence 1936–1940, 5/74, file 220, and other documents in the same location. From the correspondence in this and other files one gains the impression that the scientists, including the senior ones, enjoyed only limited academic freedom, could not collaborate with outside researchers without giving prior notice, and were required to "make up" time lost for speaking engagements and other activities outside the Institute. See there, e.g., Bergmann's

The nature and range of the research pursued at the institute during its early years can be divined from the titles of the articles published by its scientists and the types of journals in which they published their findings, as well as from the reports the institute itself issued periodically. These reveal that institute scientists appeared prominently, each of them once or twice a year, in the most important chemistry journals of the day, among them the British *Journal of the Chemical Society* and the American *Journal of the American Chemical Society*. Clearly they were engaged with the problems raised by this field's internal agenda.[173]

handwritten letters to the Sieff Institute Executive Council, on July 1, 1937, and to Bloch, for the Executive Committee, on February 14, 1939 – the latter regarding Weizmann's agreement to his teaching a course at the Hebrew University. In 1936, the Faculty of Mathematics and Natural Sciences at the Hebrew University had invited Bergmann to accept a post as Professor of Chemistry; Hebrew University Archive, meetings of the Natural Sciences Faculty, File 23, report of a meeting of the Organic Chemistry Committee, June 18, 1936. Some two years previously, Weizmann had explicitly opposed a proposal for Bergmann to accept an appointment at the university: Weizmann to Bergmann, *CWL*, XV, no. 234 (February 2, 1934). In 1936, Weizmann was elected as the first dean of the newly founded Faculty of Mathematics and Natural Sciences. Bloch's role deserves attention beyond this limited context. Certainly during the years that neither Weizmann nor Bergmann was in Rehovot, but also when they were there, Bloch seems to have been a central figure in the Sieff Institute's management. His role derived from both his scientific stature and his good nature. As with Bergmann, Bloch also continued to serve in the same capacity in the Weizmann Institute. See *Dr. Benjamin Bloch: Sketches*, Rehovot: Weizmann Institute of Science, 1960 (Hebrew). Bloch set the salaries of researchers and other employees. The correspondence with him in the Sieff Institute files at the WIA provides evidence of his treatment of the tensions at the institute. In January 1948, Bergmann, on behalf of the Weizmann Institute Planning Committee, invited Bloch to serve as head of the new institute's administrative department: Bergmann (London) to Bloch (Rehovot), January 17, 1948; WIA, 5/74/14. See also below, note 188. Bloch remained in this administrative position through the first years of the Weizmann Institute. See *The Weizmann Institute of Science*, Rehovot, November 1949, p. 56.

173 The Institute collected, bound, and dispatched offprints of its researchers' publications to professional libraries of the time. See also the list of publications appended to the institute's occasional reports. Bergmann's biographers note that from 1934 to 1939, he published seventy-eight articles, about the same number he had

The frequent contributions of Weizmann and his institute colleagues to Britain's most eminent general science publication, *Nature,* indicate that the institute was pursuing innovative research.[174] Weizmann's name stands out prominently in the list of publications, often as a co-author – for example, with Bergmann and Hirshberg. In these cases, he generally appears as the first author, with two or three collaborators.[175] Weizmann's name also appears on articles relating to industrial chemistry.[176] As noted, Weizmann stressed the importance of theoretical inquiry as well as its present or future application to agriculture and industry. The Institute's reports provide a synopsis of the substance of the articles published but do not name the researchers involved.[177]

Among the agricultural issues investigated, for example, was the exploitation of orange peels. The Weizmann organism, *Clostridium acetobutylicum Weizmann*, could be cultivated on a substrate of this material, it was shown, and would produce satisfactory quantities of acetone and butyl alcohol. Similarly, bacterial fermentation could be used on the waste products of the dairy industry to produce feedstock required by

published in Germany in 1928–1933. See Jensen et al., *Scientist in the Service of Israel* (above, note 153), p. 83.

174 In 1938–1940, Weizmann published one article each year in *Nature*; see the list of his scientific publications in this volume.

175 During his Manchester period, Fieser notes, Weizmann made a point of placing his name last on articles deriving from collaborations with his students; see Fieser, "Dr. Weizmann's Research (above, note 25). Notable among the institute's publications at this time are articles authored jointly by institute scientists and world-renowned researchers in Europe and elsewhere, e.g.: L. Farkas and E. Wigner, "Calculation of the Rate of Elementary Reactions of Light and Heavy Hydrogen," *Transactions of the Faraday Society*, 32 (1936), pp. 708–723; E. Bergmann, M. Polanyi and A.L. Szabo, "Substitution and Inversion of Configuration," *ibid.*, pp. 843–852, http://pubs.rsc.org/En/content/articlelanding/1936/tf/tf9363200843#!divAbstract (accessed December 11, 2016).

176 See, for example, no. 53 on the list of Weizmann's scientific publications in this volume.

177 Were the authors' names omitted in an attempt at collectivization, perhaps, along the lines of the Bourbaki group of mathematicians? Or was it simply an attempt to diminish Bergmann's contribution? This matter deserves further study.

local industry. In the bacteriological field, the fermentation mechanism of the Weizmann organism was investigated, as was the biochemistry of the different strains of that organism cultivated at the institute.[178] Other biochemical issues studied included the synthesis of steroids and the causes of cancer.[179] Institute scientists entered into research collaborations with colleagues at other institutions – for example, with Saul Adler, a parasitologist at the Hebrew University, for a study of chemotherapeutic treatments for the cattle parasite *Theileria*.[180]

The Sieff Institute program did not explicitly reflect Weizmann's view that all scientific forces should be mobilized to achieve certain ends, but this position can be seen in a speech he made in 1936 before supporters of the Hebrew University, where he advocated the establishment of what he termed a scientific-industrial committee.[181] Similarly, Bergmann founded a Central Committee for the Development of the Chemical Industry in Eretz Israel, which commenced work after the outbreak of World War II.[182] Outside sources indicate that, in addition to aiding civilian private industry – a declared goal of the institute – its scientists made chemical and industrial contributions, in the early

178 For a comprehensive survey of the possibilities for the development of the fermentation industry and the special role played by Palestine/Israel in this field, see Ernst David Bergmann, "Utilization of Agricultural Products," in J.B. Hobman (ed.), *Palestine Economic Future: A Review of Progress and Prospects*, London: Percy Lund Humphries, 1946, pp. 130–136.

179 See, e.g., Ernst Bergmann, "Stearins, Sexual Hormones and Similar Substances," *Tekhnikah umada'*, 1 (1937), pp. 9–11 (Hebrew).

180 "The Daniel Sieff Research Institute," *Tekhnikah umada'* 2/8 (above, note 133), p. 24; WIA, file 439-1249. Compare this with the limitations on collaborations described in note 172. On *Theileria* and Adler's work, see Rivka Ashbel, *As Much As We Could Do: The Contributions Made by the Hebrew University and Jewish Doctors and Scientists from Palestine during and after World War II*, Jerusalem: Magnes, 1986, pp. 57–60 (Hebrew).

181 See *The Mountain and the Valley*, Tel Aviv: Committee of the Friends of the Hebrew University of Jerusalem in Tel Aviv, 1936, p. 34 (Hebrew).

182 Itzhak Gilon-Zielinski, *The History of the Chemical Industry in Israel*, Tel Aviv: Committee for Book Publication, 1976, pp. 18–20 (Hebrew). See Orna Zeltzer, "Science and the Yishuv," paper for archival studies program, Beit Berl, 2008 (Hebrew; personal copy).

stages of the Yishuv's military industries, to the production of explosives for the Haganah. The best-known of these projects was the involvement of the brothers Felix and Ernst Bergmann in the production of "TEN," pentaerythritol tetranitrate, at the end of the 1930s.[183]

From time to time, prominent European scientists lectured and offered seminars at the institute.[184] Weizmann's desire, explicit in the plan, that it serve as a haven for young scientists who had completed their studies and sought advanced training in their profession, was realized when it began to take in research students from the Hebrew University. These had apparently chosen institute scientists as mentors in addition to their tutors on Mount Scopus, or had been attracted by the more advanced

183 See Uriel Bachrach, *By the Power of Knowledge: The Scientific Corps of the IDF (1948–1952)*, [Tel Aviv]: N.D.D., 2009, p. 27 (Hebrew). Bachrach dates such activity to 1939 and places its base in the Jafora plant in Rehovot. See also Joseph Evron (Abbo), "Products and Development ('TEN')," in *The Israel Defense Industry*, Tel Aviv: Ministry of Defense, 1980, pp. 29–35 (Hebrew). Truly insider evidence comes from Shalom Sarel, who would later become a pharmaceutical chemist at the Hebrew University: "In addition, and in secret, apparently without Chaim Weizmann's knowledge, the two Bergmann brothers – the elder, Ernst, and the younger, Felix – contributed to a war effort of a different type: aiding the Yishuv in the development of the state-in-the-making's military-industrial infrastructure. Felix Bergmann, with typical discretion, provided the Haganah with valuable advice in a variety of areas. This was kept secret to avoid embarrassing Dr. Weizmann, who had, over the years, developed relations of trust with the British Mandate administration." Idem, "Memories from the Scientific Revolution," http://heb.wis-wander.weizmann.ac.il/אבן-דרך/זכרונות-מהמהפכה-המדעית (Hebrew; accessed December 12, 2016). See also Munya M. Mardor, *Rafael: In the Paths of Research and Development for Israel's Security*, Tel Aviv: Ministry of Defense, 1981, p. 93 (Hebrew).

184 Such as Jacques Errera from Brussels, Victor Henri from Liège, Carl Neuberg, and Carl Oppenheimer, both from Berlin. See "The Daniel Sieff Research Institute," *Tekhnikah umada'* 2/8 (above, note 133), p. 14. According to Weizmann, these visits were aimed at establishing personal connections with the larger scientific world and were thus of special importance for a new institute in a country physically distant from the major centers of scientific research. The importance of such activities was already highlighted in the 1914 plan.

equipment available in Rehovot, or both.[185] Weizmann himself thought that the working conditions at the institute were "much more comfortable and convenient than in many great academic institutions which I had the opportunity to see, and I have seen the University of Manchester [Lister Institute], the laboratories in South Kensington, in Oxford, in the Sorbonne, and in the Pasteur Institute."[186]

The Sieff Institute entered into a very different period as a new world war threatened and then broke out. Weizmann's agenda changed, and his national leadership role kept him far from Rehovot for long periods. Nevertheless, his responsibility for the Institute led him to embark on two initiatives. First, he established a new source of funding for the institute in the United States. Second, he furthered research projects aimed at industrial needs and directly connected to what Weizmann saw as the war effort. A byproduct of this latter move was the potential for successful projects to serve as additional sources of funding for the institute. Such projects were a component of Weizmann's plans for the Yishuv's

185 Among the first Hebrew University doctoral students at the institute were Eliahu Bograchov, Benjamin Elazari-Volcani, and Tuvia Berlin. Bograchov's research topic was "Experiments in Connection with the Preparation of New Carcinogenic Materials and an Examination of their Structures," submitted to the Hebrew University in 1939, http://lib-phds1.weizmann.ac.il/Dissertations/BOGRACHOV_ELIAHU.pdf (Hebrew). His adviser in Jerusalem was Moshe Weizmann, while Ernst Bergmann supervised him in Rehovot. Benjamin Elazari-Volcani's work was "A Study of Life in the Dead Sea," submitted to the university in 1940, but containing results that had already been published in *Nature* in 1936; see Aharon Oren & Antonio Ventosa, "Benjamin Elazari Volcani (1915–1999): Sixty-Three Years of Studies of the Microbiology of the Dead Sea," www.im.microbios.org/07setember99/11%20Oren%20(P).pdf. His adviser in Jerusalem was Israel Jacob Kligler; in Rehovot Chaim Weizmann himself oversaw his work. Tuvia Berlin investigated polycyclic aromatic compounds, working under Moshe Weizmann in Jerusalem and Ernst Bergmann in Rehovot, http://lib-phds1.weizmann.ac.il/Dissertations/Berlin_Tuvia.pdf (Hebrew, all accessed December 12, 2016). The Sieff Institute operated as a kind of research extension of the Hebrew University, which imbued the connections between the two with great significance. In other words, they were not simply hostile rivals.

186 Weizmann to Weisgal, *CWL*, XIII, no. 325 (Aug. 18, 1949).

participation in the war effort, which centered on the enlistment of Palestine's Jews in the British army. In a memorandum Weizmann sent to figures in the British government, he noted three relevant areas to which the Yishuv could contribute: production of chemicals, provision of mechanical repair facilities, and production of medical supplies.[187] Palestine Pharmaceutical Products Ltd. (PPP) was one initiative that arose out of the Sieff Institute with the intention of exploiting its research capabilities, particularly in pharmaceuticals.[188] Another example was the utilization of the institute's expertise in fermentation processes to exploit local raw materials. The difficulty of exporting citrus fruit made it a convenient raw material, for which the Sieff Institute scientists found innovative uses, such as the production of anti-knock engine additives.[189]

187 [Weizmann], "Palestine's Possible Contribution to the War Effort," July 22, 1940, WIA, 14-2240A. On the context see Weizmann to Lord Lloyd, *CWL*, XX, no. 31 (August 25, 1940), in particular the editors' clarification.

188 Weizmann informed his colleagues on the institute's Executive Committee that the appropriate arrangements had been made regarding the institute's legal standing so that PPP would be able to manufacture limited quantities of drugs and serve as a pioneer for large-scale production; see Weizmann to Isaac Magasanik, Harold Davies, and Frieda Goldschmidt, *CWL*, XX, no. 18 (July 26, 1940), pp. 22–24. Weizmann sought to enlist Neuberg in this initiative and to make him responsible for the bacteriological and biochemical work that the Sieff Institute would do for the company; see [PPP], Minutes of the Meeting of July 24th, 1940, CWA, 5/74, file 16, Sieff Institute Board of Directors. Bloch was responsible for the initiative, which was set up on the institute's premises. "This company was the first to provide, in Palestine and to the Allied armies, important medical preparations synthesized at all stages on the institute premises"; Yehuda Hirshberg, "Reminiscences," in *Bloch: Sketches* (above, note 172). PPP established a partnership with the Green Brothers company for the production of, among other things, Phanodorm (a sleeping aid); an alternative to the anti-malarial drug Atabrine; and Ichthyol, a treatment for skin ailments. The company operated from 1939 to 1949. See Amnon Mikhlin, *Pharmacists' Ventures: The History of Pharmacy in Eretz Israel*, Tel Aviv: Ministry of Defense, 1999, p. 224 (Hebrew). See also Gilon-Zielinski, *History of the Chemical Industry* (above, note 182), pp. 86–87, 96. On the research aspect, see Adi Sapir, "Commercialization of Academic Science in Israel," Ph.D. dissertation, Tel Aviv University, 2014, pp. 113–117 (Hebrew).

189 On Bergmann's involvement see Ernst Bergmann, "The Use of Citrus Fruit" (July

In 1946, marking its first decade, Bergmann asserted that the Sieff Institute was continuing to carry out Weizmann's original concept of commitment to basic research, in the belief that it would eventually lead to results that would respond to the country's needs.[190]

Weizmann himself combined his political activity in Great Britain, the United States, and Palestine with his own research initiatives, including a new version of the synthetic rubber project.[191] Another project aimed at overcoming the shortage of essential raw materials was undertaken jointly by Weizmann and Bergmann: the conversion of the crude residues of oil distillation into aromatic substances such as benzine and

1939), a six-page typewritten memorandum intended for "Cooperation between the Scientific and Technological Institutions in Eretz Israel," WIA, (512) 11-2002 (Hebrew).

190 Bergmann, "The Daniel Sieff Research Institute" (above, note 159). In 1946, just prior to the transition of the Daniel Sieff Institute into the Weizmann Institute, Bloch summed up the research conducted by the Sieff Institute in its role as the nucleus of the new institution. He cited in particular the study of diseases, such as leprosy and tuberculosis; the study of the synthesis of chemotherapeutic compounds, including anti-malarial drugs and sulfas; and biochemical research on microorganisms that could be exploited in fermentation studies. Bloch viewed all this as pure research, noting also the applied research conducted at the institute, which formed the basis for the synthetic production of medicines of a quality equal to those produced in the past in Germany. See his article in *Palestine and the Middle East*, 18/6 (1946), p. 131. Over twenty years later, Sieff Institute researcher Ernst Simon wrote a concise survey of the research done there: "Dr. Chaim Weizmann as a Scientist," *Mada'*, 13 (July–October 1968), pp. 80–81 (Hebrew).

191 A description of Weizmann's life and political activity during this period can be found in Rose, *Chaim Weizmann* (above, note 56). In July 1940, British Supply Minister Herbert Morrison appointed Weizmann honorary chemical adviser, providing him and his assistants with a laboratory in London at 25 Grosvenor Crescent Mews. See also Weizmann to the manager, Alexandra Hotel, London, *CWL*, XX, no. 12 (July 23, 1940). A survey of Weizmann's initiatives to use his ideas in the war effort can be found in "Scientific Activities during World War II," Memorandum for Mountbatten, July 27, 1947, *CWP*, II, no. 89, pp. 652–656. A comprehensive account of Weizmann's research in the United States and Britain is offered by Bergmann in *Mada'* 17/3 (August–September 1972), pp. 169–175 (Hebrew).

toluene – the basis for TNT.[192] Beyond the war's impact on the institute's research program, however, concerns about sources of funding prompted a reconsideration of its organizational character and mission. This began with a plan for a large expansion and ended with the Sieff Institute's transformation from a small institution with a limited purview into a large multidisciplinary facility, a scientific campus, today's Weizmann Institute of Science.[193]

The Sieff Institute's dozen researchers and their laboratories represented a considerable addition to the Yishuv's scientific resources and research capabilities in the 1930s and 1940s; they played an important role in the war effort and provided a welcome scientific enhancement for ten or so Hebrew University research students.[194] In historical perspective, however, the importance of the Sieff Institute lies in serving as the precursor to the Weizmann Institute of Science[195] and in providing Ernst Bergmann with an opportunity to train as a research director and, as will become evident below, to fashion and execute an independent science policy of his own.

192 See *Trial and Error*, p. 547. One source of information for Weizmann's and Bergmann's work in this field is the patents that they registered at the time. [See the list of Weizmann's patents in this volume. – Ed.]

193 The vision for the Weizmann Institute of Science included a significant expansion of the Sieff Institute campus. The planners, including the project architect, Arieh El-Hanani, used the term "city of science," a reference to the expanded campus and not, as some have mistakenly interpreted it, an indication that it was to be a settlement of a novel sort that would transform Rehovot from an agricultural settlement into a science-based urban center. The cover of *Palestine and the Middle East*, 18/6 (above, note 190) shows the architectural mock-up of the planned campus with the caption "City of Science: Model of Future Weizmann Institute of Science at Rehovoth." The journal's feature article by El-Hanani makes it clear that "city" referred to the campus; *ibid.*, p. 130. Weizmann himself, in the issue's opening article, refers to it as the "Rehovoth Research Centre"; *ibid.*, p. 125.

194 See note 185 above.

195 In 1952 the Sieff Institute still existed formally within the Weizmann Institute; see Daniel Sieff Research Institute, "Minutes of the Annual General Meeting of the Association," September 23, 1952, WIA, 70.

From the Sieff Institute to the Weizmann Institute

In 1944, in honor of Weizmann's upcoming seventieth birthday, his long-time admirer and ally Meyer Weisgal (1894–1977) proposed that Weizmann's American friends support the expansion of the Sieff Institute through a $1 million donation.[196] The envisioned multidisciplinary research center was to be named after Weizmann.[197] That Weizmann himself envisioned such an institution is evident from a letter he wrote to Einstein during the Sieff Institute's first year of operation. Considering the initiative's success, he suggested to Einstein the establishment of another four or five similar research institutes in Rehovot, which could serve as an alternative to the Hebrew University of Jerusalem.[198]

But the program encapsulated in Weizmann's letter to Einstein was not merely a product of his disappointment at the dreams left unfulfilled by the Hebrew University. The roots of what became the interdisciplinary institution in Rehovot were put down in 1939, as the clouds of war

196 See Weisgal, *So Far* (above, note 157), p. 214.

197 *Ibid.*; Weisgal's initiative is mentioned at the year 1944 in the chronological table in *The Odyssey of an Optimist: Meyer W. Weisgal – An Anthology by His Contemporaries*, New York: Atheneum, 1967, p. 262. The early years of the Weizmann Institute, from its inception through its first years of activity, appear in the table as well.

198 Weizmann to Einstein, *CWL*, XVI, no. 316 (July 12, 1934). Weizmann proposed a scientific committee comprising Einstein, James Franck, Max Born, Richard Willstätter, Lord Rutherford, Herbert Samuel, and Robert Cecil. He suggested institutes for genetics, applied botany, pharmacology, physiology, pathology, and other fields, indicating a departure from his earlier views on science as he turned more toward modern biological research. The proposed areas of research extend beyond the study of the local natural environment but remain potentially connected to Weizmann's industrial biotechnology project and to medical research. Whether his motivation was competition with the Hebrew University or a fundamental change in his thinking about science and science policy is a matter for further inquiry. Clearly, in his planning for both the Sieff and the Weizmann Institutes, he displayed a clear preference for the spectrum of physical chemistry fields. An editorial note to this letter indicates that Einstein opposed the idea, for economic reasons.

were gathering.[199] Weizmann asked Josef Cohn (his aide in 1935–1938) to travel to the United States to put together a basis that would make it possible to raise funds for the Sieff Institute there, in addition or as an alternative to fundraising in Britain. In 1940, the Chaim Weizmann Research Foundation was registered in the U.S., its stated goal being to promote chemistry research in Palestine. In 1944 it was renamed the American Committee for the Weizmann Institute of Science.[200] Weisgal obtained the agreement of the Sieff and Marks families to integrate the Sieff Institute into a larger organizational framework to be called the Weizmann Center. After further thought, it was decided that the name would be the Weizmann Institute of Science.[201]

The cornerstone for the new center was laid in April 1946, in a ceremony accompanied by an international scientific symposium. Some of the leading chemists and biochemists of the era participated, along with some of the Yishuv's most important scientific figures, all in honor of Weizmann and the new institute that would bear his name. The proceedings were published under the somewhat grandiose title *Trends in Modern Science* and were meant to mark out a clear direction in academic policy.[202] This seems to have been the first international scientific conference ever held in Palestine.[203] That year, in addition to

199 Flowers (above, note 129), p. 48; *Odyssey of an Optimist* (above, note 197), pp. 165–166.

200 Flowers (above, note 129), p. 50. See also *Celebrating 70 Years of Science and Philanthropy*, a film produced to mark the committee's seventieth anniversary, https://www.youtube.com/user/weizmannusa (accessed December 13, 2016).

201 In January 1943, a plan to expand the Sieff Institute had been drawn up in response to the growth of local industry during the war. In terms of its future research program, however, this expansion adhered to the original concept on which the Sieff Institute had been founded ten years previously; see "The Future Development of the Sieff Institute," WIA, (96) 6–74. The plan for the Weizmann Institute was based on an entirely different scientific and organizational vision.

202 *Trends in Modern Science: Symposium Held at Rehovot on the Occasion of the Laying of the Cornerstone for the Institute of Physics and Physical Chemistry*, Rehovot: Weizmann Institute of Science, 1949. A summary of the talks can be found in *Palestine and the Middle East*, 18/6 (1946), p. 134.

203 See Weizmann's introduction to *Trends in Modern Science* (above, note 74). A year

the existing departments already functioning at the Sieff Institute, the incipient Weizmann Institute was scheduled to acquire new research departments in fields at the scientific frontier of the day: applied mathematics, isotope research, biophysics and physical chemistry, optics, and high polymers.[204]

About three years passed until the official opening of the Weizmann Institute, which was held on November 2, 1949. It was a symbolic date for Weizmann – the anniversary of the Balfour Declaration. At the ceremony, the building housing the Institute of Physics and the Institute of Physical Chemistry was also dedicated. That same year, Weizmann had been chosen to serve as president of the State of Israel, and subsequently he became honorary president of the Weizmann Institute.

Weizmann's failing health limited his involvement in the new institute's affairs. He died three years later, on November 9, 1952, and Meyer Weisgal, chairman of the Executive Council and Board of Governors, found himself at the head of the institute and responsible for Weizmann's legacy, including the publication of his writings. By 1953, the institute's organizational structure was in place, with Weisgal at its head. Serving under him was a scientific director elected by the institute's Scientific Council, which, replacing the temporary founding Scientific Council headed by Herman F. Mark (1895–1992), was made up of the senior scientists who headed the institute's central units: Isaac

later, in the summer of 1947, the first international conference on Jewish studies was held on Mount Scopus. See *The First World Conference on Jewish Studies*, Jerusalem: Magnes, 1962 (Hebrew).

204 Ernst D. Bergmann, "New Home for Science," in *Palestine and the Middle East*, 18/6 (1946), pp. 129, 133. See also the Sieff Institute "Governors' Report of the Activities of the Institute during the Year 1946," which includes a budgetary report, in WIA (no reference numbers); and see the booklet *The Weizmann Institute of Science* (above, note 172). This booklet, published in November 1949 in advance of the dedication ceremony, presents Weizmann as president of the Executive Council, Weisgal as its chairman, and Bergmann as scientific director; *ibid.*, p. 53. Herman Mark is presented as chairman of the Planning Committee and James Franck as chairman of the institute's Advisory Committee. Einstein, Robert Oppenheimer, and John von Neumann are listed as members of the Advisory Committee of the Department of Applied Mathematics; *ibid.*, p. 57.

Berenblum, experimental biology; Israel Dostrovsky, isotopes; Ephraim Katchalsky-Katzir, biophysics; and Chaim Leib Pekeris, applied mathematics. Aharon Katchalsky-Katzir, head of the polymers department, chaired the council.[205] This structure kept fundraising and management separate from academic administration, and placed academic direction in the hands of the institute's scientists. Where, then, was Bergmann?

Israel's War of Independence had intervened between the cornerstone-laying and the official opening ceremony. In the spring of 1948, over a score of Jerusalem-based scientists from the Science Corps of the Haganah, soon to become the Israel Defense Forces, gathered in the compound that housed the Sieff Institute, the Agricultural Experimental Station, and the Hebrew University Faculty of Agriculture; they would later be joined by several dozen more scientists from other locations.[206] Matters developed such that Bergmann, scientific director of the Sieff Institute and scientific director-designate of the Weizmann Institute, became the authority for the Science Corps' personnel and their work.[207]

205 *The Weizmann Institute of Science, Report for 1953*, Rehovot, November 1953, p. 97. Note that the projects of these central departments do not include any of Weizmann's biotechnological initiatives. His microbiological research was, however, carried on by Esther Hellinger, one of the first researchers at the Sieff Institute. The Scientific Council under Aharon Katchalsky-Katzir seems to have been a continuation of the one that had been headed by Bergmann.

206 The Haganah command officially formed the Science Corps on March 17. See Ephraim Katzir, "The Beginnings of Defense Research," in Joshua Jortner (ed.), *David Ben-Gurion and the Development of Science in Israel*, Jerusalem: Israel Academy of Sciences and Humanities, 1989, pp. 25–49 (Hebrew). See also Katzir, *Life Story* (above, note 124), pp. 100–107.

207 This is attested in a document dated May 23, 1948, a kind of "order of the day," which Bergmann sent to the Institute's personnel regarding procedures during this time of emergency; WIA, 14-5/74. Bergmann signed the document without adding any title or military rank. In his two-page CV dated February 22, 1952 (above, note 153), four years after these events, Bergmann wrote: "From the opening of the Sieff Institute in April 1934 and until July 1951, I was director of the institute, and from November 1949, I was also the director of the Weizmann Institute of Science. In July 1951 I left for a year's sabbatical, which was extended for another year at the insistence of the defense minister [Ben-Gurion]. During this period I devoted myself to Ministry of Defense research and planning, which I head. I have for all

Bergmann thus reached the pinnacle of the Science Corps command by virtue of his position at the Sieff-Weizmann Institute at the outbreak of Israel's War of Independence. During these weeks, the Sieff-Weizmann compound was converted to house Science Corps personnel and others from similar units working on armaments. Only later was Bergmann given an official Defense Ministry appointment. In any event, for Bergmann, managing the Science Corps base in Rehovot seems to have been more than just an organizational event – it involved a much more profound change in his views, as can be seen in a letter he wrote to Weizmann in May 1948: "Sometimes I wonder if we were right to devote fourteen years to the type of research we did instead of preparing the country, chemically and technically, for the sophistication of modern warfare."[208]

Bergmann, an avid supporter of Weizmann's vision and the person who, more than any other, was responsible for implementing the Sieff-Weizmann Institute's development program, did not simply modify his thinking. He renounced his previous views, adopting instead an alternative vision of a national research institute that would bring together, under one organizational roof, all of the government's research units. This entity would effectively replace both the Weizmann Institute and the Science Corps. At first, Bergmann tried to persuade Weizmann, in his capacity as president of Israel's Provisional State Council, to adopt this initiative. Weizmann rejected it out of hand and said as much at the dedication of the Weizmann Institute.[209] Alongside mentioning, with

intents and purposes held this position since 1948. In 1951 I was also appointed scientific adviser to the defense minister." See also Arieh Boaz, *The Origins of the Ministry of Defense*, Tel Aviv: Modan–Ministry of Defense, 2013, p. 158 (Hebrew).

208 Bergmann (Rehovot) to Weizmann (Switzerland), CWA, 17-2839 (May 31, 1948). Nothing more strongly demonstrates Bergmann's allegiance to Weizmann's vision and science policy, and the profound contrast between it and Bergmann's own new views, than his 1946 article "Utilization of Agricultural Products" (above, note 178).

209 After the official ceremony, Weizmann made Weisgal promise never to abandon the Weizmann Institute and to devote the rest of his life to it; see Weisgal, *So Far*

modest satisfaction, the participation of the institute's researchers in the war effort, Weizmann appealed directly to Bergmann, praising his talents as a scientist, and named him the director of the new institute. However, before symbolically handing Bergmann a set of gold keys, he told him: "I am confident that you, and the staff, will maintain the purity of science, for the sake of our country and the sake of all of humankind."[210]

Bergmann, however, did not accede to the plea of the man who had for so long seen in him a loyal ally and almost a close family member; he no longer labored to ensure that the Weizmann Institute would be a citadel of pure research. Bergmann had many supporters among researchers affiliated with the institute or working in its facilities, while Weizmann was backed by the members of his close circle, chief among them Weisgal and the leaders of the Weizmann Institute project, American Jews. The rift between the two men widened. For a long time the institute seethed. From within, Weisgal attempted to mend relations,[211] and from outside it, at Ben-Gurion's behest, former IDF Chief of Staff Lieutenant-General (ret.) Yaacov Dori (among others) tried to bridge the gap between the two, but to no avail.[212]

The denouement would have been inconceivable just a short time earlier: Weizmann, who saw the institute bearing his name as his personal legacy, fired Bergmann. Weisgal drafted the dismissal letter and

(above, note 157), p. 273. Does this appeal to Weisgal mean that Weizmann had begun to question Bergmann's loyalty to the mission and legacy that Weizmann had imposed on him?

210 The CWA holds a document entitled "Address of President Weizmann" that is not included in either *CWL* or *CWP*, as well as a Hebrew version of Weizmann's speech and clippings of reports on it in Hebrew newspapers. On the opening ceremony, see Rose, *Chaim Weizmann* (above, note 56), p. 284.

211 See Weisgal's letter to Dewey Stone and Harry Levin, July 8, 1951, WIA, 6-2928. I will expand on this incident in my article, "Bergmann and the Science Corps: To What Extent Can Weizmann Be Credited with Establishing a Fourth Research Institute," in preparation.

212 At that time, Dori had been charged by Ben-Gurion with establishing a body to oversee the country's higher education and science institutions. See Mordecai Naor, *Yaacov Dori: IDF First Chief of* Staff, Tel Aviv: Modan 2011(Hebrew).

Weizmann signed it on July 2, 1951.[213] As Weizmann's health progressively declined, he could see from his home in Rehovot the Weizmann Institute once more becoming an embodiment of his academic policy. It was now led by his confidant Weisgal, who identified fully with Weizmann's ideas, and its departments were headed by scientists of international stature who served also as the institute's Scientific Council. Alongside their research at the institute, and with varying levels of commitment, these figures devoted their abilities and prestige to work relating to Israel's security, thereby creating work patterns that clearly distinguished between their contributions to academic and military research.[214] After his dismissal, Bergmann was officially appointed science adviser to the defense minister.[215]

From Weizmann's perspective, the initial organizational planning and fundraising stages of the Hebrew University, the Sieff Institute, and the Weizmann Institute differed from one another fundamentally. Weizmann had a number of partners in the Hebrew University project; he had not chosen them, and not all of them agreed with his view of what that institution's academic policy ought to be. As a result, not all went as he wished in its early years. The opposite was the case with the Sieff Institute. There, he was the sole determiner of academic and research policy. He also led the institute's fundraising and day-to-day operations and chose its research staff. By contrast, in the case of the Weizmann Institute, which was meant to embody his academic and research policy, he was not the executor of the concept. The preparatory

213 Weizmann to Bergmann, *CWL*, XXIII, no. 349 (July 2, 1951), p. 304: "After due consideration, I have come to the decision to relieve you of your duties and responsibilities as Scientific Director of the Weizmann Institute. You will continue your functions as Head of the Department of Organic Chemistry. It is, however, understood that you will take your Sabbatical leave as from the 15th July, 1951." See also the editorial notes appended to the letter. The letter of dismissal was one of the last dozen or so letters that the ailing Weizmann wrote.

214 See Katzir, *Life Story* (above, note 124), p. 114: "a kind of clear separation of powers became the rule at the institute, beginning, so it seems, on November 2, 1949."

215 See Bergmann's CV (above, note 153).

work was done principally by his allies Weisgal and Bergmann, who did all they could to fulfill his vision. Weisgal used his connections with wealthy American Jews and community leaders to provide a financial base for the project. As in the case of the Sieff Institute, the guarantee of financial support preceded planning and actual establishment. The donors, members of the American Committee for the Weizmann Institute of Science, maintained from the start that the institution required an endowment of $1 million.[216]

Like Weizmann, Bergmann spent a lot of time in the United States in those years, and while there he addressed the academic aspects of the project. In the past, it had been Weizmann's connections with key scientific figures that helped him recruit researchers, especially junior ones; it was now up to Bergmann to make use of his own connections. Chief among them was Herman Mark, whom he successfully recruited to head the academic planning committee for the new institute. Mark was a celebrated organic chemist of the first order who had done pioneering work on large molecules and laid the theoretical groundwork for this new field of study.[217] Bergmann had first made Mark's acquaintance

216 Weisgal, *So Far* (above, note 157), p. 214. On the Sieff Institute's budget from the time of its founding, the transition during the war to dependence on fundraising and support in the United States, and Weizmann's request that the Sieff and Marks families continue to support the original institution as a component of the evolving Weizmann Institute, see Weizmann to Sieff, *CWL*, XXII, no. 15 (April 24, 1946), pp. 131–134. For details of the budget of nearly $1 million, see "American Committee for the Weizmann Institute of Science, Science Builder & Beacon, Budget for 1948," p. 7, WIA, Weisgal file, reports and publications box.

217 Herbert Morawetz, *Herman Francis Mark, 1895–1992: A Biographical Memoir*, Washington: National Academies Press, 1995, pp. 193–205; www.xtal.iqfr.csic.es/Cristalografia/archivos_10/mark-h-f.pdf (accessed December 13, 2016). Mark's autobiography devotes considerable space to his role in helping found the Weizmann Institute; see Herman Mark, *From Small Organic Molecules to Large: A Century of Progress*, Washington: American Chemical Society, 1993. On future Israeli President Ephraim Katzir's connection to Mark's research, which began during Katzir's student days, see Katzir, *Life Story* (above, note 124). The basic textbook by Kurt Meyer and Herman Mark on large natural organic molecules

at the University of Berlin, in Wilhelm Schlenk's laboratory.[218] Once Mark had accepted Bergmann's invitation, he appointed subcommittees to lay out the preferred research fields and identify appropriate candidates for appointments at the Weizmann Institute. He carried this out from the United States, where he enlisted leading scientists to head these subcommittees, among them Robert Oppenheimer, who led the group on applied mathematics, and James Franck, who led the group on physics and physical chemistry.[219]

These subcommittees considered which notable scientists would be

was a fixture in the Mount Scopus laboratory of Katzir's teacher, Max Frankel; *ibid.*, p. 59.

218 Bergmann wrote one of his first studies on isomers together with Mark; see the list of Bergmann's publications in Jensen et al., *Scientist in the Service of Israel* (above, note 153), p. 301, no. 4 (1929). The WIA contains a letter written by Mark to Bergmann in April 1930, and another from Mark (then at the University of Vienna) to Bergmann (erroneously addressed to the "University of Jerusalem") dated June 25, 1937, attesting to an ongoing relationship between the two; see WIA, 5/74(2). I am grateful to Pnina Abir-Am for suggesting that I seek out the origins of the Bergmann–Mark relationship in Berlin.

219 Weisgal, *So Far* (above, note 157), p. 225. Weizmann took part in fundraising events and no less so in shaping the plans and priorities of the institute's research program. He pushed for a focus on physical chemistry, biochemistry, and pharmacology; regarding organic chemistry, he maintained that the existing Sieff Institute program was sufficient. See Weizmann to Mark, *CWL*, XX, no. 9 (May 17, 1945), pp. 9–10. Among the members of the American Committee for the Weizmann Institute of Science were the chemist W.P. Hohenstein and the biochemist M.D. Kamen. The latter, like Oppenheimer and Franck, was one of the senior scientists working on the Manhattan Project; see Weizmann to Weisgal, *CWL*, XX, no. 322 (April 28, 1947), notes 3 and 5. Kurt Stern served as secretary of the committee and was himself designated to head the physical biochemistry department. For the working plans of this department, see Kurt G. Stern, "Department of Physical Biochemistry," in *Palestine and the Middle East*, 18/6 (1946), p. 133. To the British Committee for the Weizmann Institute, formed by British scientists Ernst Chain and Isaac Berenblum, Weizmann suggested that a biochemistry and biological institute be established; see Weizmann to Mark, *CWL*, XX, no. 368 (July 15, 1947). In some of the plans for the Weizmann Institute, fermentation research was to be integrated into an institute of biology and pharmacology, of which Chain and Berenblum would be part; memorandum dated October 16, 1946, WIA, 6/74(23).

asked to join the planned research units of the Weizmann Institute. At the same time, preparations were made to purchase the requisite scientific equipment. This integrated strategy was exemplified by the recruitment of applied mathematician Chaim Leib Pekeris for the institute, concomitantly with the construction of WEIZAC (the Weizmann Automatic Computer) to be used as a research tool by him and others. WEIZAC was Israel's first computer, and one of the first in the world. It was built under the leadership of John von Neumann on the model of the one at the Institute for Advanced Study at Princeton. Bergmann recruited Pekeris.[220] Mark was a much more successful chairman of the Scientific Committee than Einstein had been as head of the Hebrew University's Academic Council some two decades earlier. Mark's success raises questions regarding the wisdom of Weizmann's decision to offer Einstein that appointment.[221]

The new institute did not meet all expectations. Not all the scientists that Mark's committee solicited actually arrived. In contrast to the early days of the Hebrew University, however, alternative younger candidates were already available among scholars born and educated in Israel. The institute's research units as of 1953 were: applied mathematics, headed by Pekeris; biophysics, headed by Ephraim Katchalsky-Katzir; experimental biology, headed by Berenblum; isotopes, headed by Dostrovsky; optics, including an infrared laboratory, headed by Joseph Henry (Joe) Jaffe; photochemistry, headed by Hirshberg; x-ray crystallography, headed by Gerhard Martin Julius Schmidt; polymers,

Fermentation and microbiology come under biochemistry in the 1948 budget memorandum (above, note 216), p. 20.

220 For the history of WEIZAC see Raya Leviathan, "One State, One Computer: Building the Weizac Computer at the Weizmann Institute in the Mid-Fifties, The Decision and its Effect," Ph.D. Dissertation, Tel Aviv University, 2015 (Hebrew). Bergmann expected that "the Department of Applied Mathematics will be the brain to which theoretical problems encountered by the other departments will be referred"; Bergmann, "New Home for Science" (above, note 204), p. 129.

221 There has yet to be an evenhanded study of the Hebrew University's history that takes up the question of Einstein's inefficacy. See again Weizmann to Einstein, *CWL*, XV, no. 409 (June 8, 1933).

headed by Aharon Katchalsky-Katzir, who was also chairman of the institute's Scientific Committee; and organic chemistry, headed, temporarily, by Ephraim Katchalsky-Katzir. As noted above, Bergmann, who had served at Weizmann's behest as scientific director, was not among the new institute's unit heads, nor even among its scientists.[222] At that time he was involved in establishing the organizational infrastructure for military research and its scientific management in Israel, following his failure to convert the entire Weizmann Institute of Science into an organization oriented toward military research, which would have differed drastically from the vision of its founders – including Bergmann himself.[223]

In the final analysis, if the Sieff Institute was meant to enable Weizmann and his partners to operate in keeping with Weizmann's view of science, the Weizmann Institute was meant to realize Weizmann's original vision for the Hebrew University. It was the vision he had formulated at the beginning of the 1920s, apparently inspired by his encounter with Paul Ehrlich and the drafting of a plan for a medical research institute in Jerusalem. The new entity was meant to be an umbrella under

222 Weizmann Institute report, 1953 (above, note 205). The new institute had several other units as well: Plant genetics was headed by Oved Shifriss; microbiology, including its industrial applications, by Esther Hellinger; and microbial metabolism by Benjamin Elazari Volcani. In his *Life Story* (above, note 124), in the chapter "Between Weizmann and Weisgal," Katzir lists those nominated to head the departments, among them David Rittenberg, Peter Hohenstein, and Kurt Stern, "who decided to pass up the honor and the trouble (and what must have seemed to them, most likely, as mortal danger) and to remain in their laboratories in New York." Katzir adds: "There was nothing to do but to appoint 'local' replacements who were originally meant to serve as their assistants" (p. 109).

223 Nevertheless, a month after Weizmann's death, Bergmann assumed the role of featured speaker at a conference held in his memory, which aimed to appraise Weizmann's scientific work. He used the opportunity to extol the Science Corps and atomic energy, the projects he had worked on at Ben-Gurion's behest. Ernst Bergmann, "Principles of Chemical Synthesis in the Laboratory and in the Living Cell," address delivered to the Israel Chemical Society, December 15, 1952, WIA, Bergmann file (Hebrew). See also Bergmann on Weizmann, in Bergmann, "Obituary Notice" (above, note 38).

which a number of research institutes would operate. The researchers would not be obligated to teach, and their research purviews would be determined by the fields of the world-renowned scientists who would come to Rehovot. Furthermore, the new version of Weizmann's science policy privileged experimental and mathematical research, which at the time were seen to be at the forefront of science. The fundraising mechanism set up by Meyer Weisgal guaranteed the construction of sophisticated laboratories and the purchase of state-of-the-art scientific equipment for the scientists who joined the new institute. After Weizmann's death, when Weisgal was appointed to lead the institute, he saw it as his obligation to bring Weizmann's concept to fruition. Weisgal cited his mentor: "Take your standards elsewhere, from Princeton, Cambridge, Oxford, MIT, Copenhagen. These are the centers of learning we must emulate."[224] At the same time, the translation of this vision into everyday scientific activity was incumbent upon the scientists themselves and the Institute's autonomous academic bodies.

That the Hebrew University, the Sieff Institute, and the Weizmann Institute of Science were among the first research organizations in the Yishuv and the early years of statehood raises the issue of the "founders' effect." How and to what extent were Weizmann's scientific policy and its organizational expressions transmitted to generations of researchers and research institutes who did not "know" them? How did Weizmann's legacy mold Israel's academic and research system in the decades that followed? Another question, no less important, has to do with Bergmann and his role as the founder of a powerful state military research complex. To what extent was Bergmann successful in functioning within this new system as a bearer of Weizmann's vision of the place of basic research, and in integrating that vision into an organization focused on applied military projects? Munya Mardor, Bergmann's partner in setting up the defense research system, remarked in the 1950s: "We were aware of the fact that pure research was a value in

224 Weisgal, *So Far* (above, note 157), p. 281. Weisgal referred to himself as the head of the Weizmann Institute no matter what specific title he actually held; *ibid.*, p. 302.

and of itself, and that even an organizational entity involved in military research and development had to view basic science as vital and include it in its work programs."[225]

At the beginning of the 1950s, the Hebrew Technion officially changed its name to the Technion – Israel Institute of Technology, acknowledging its status as a technological research university and demonstrating that not all the components of the Israeli research system owe their origins to Weizmann. Nevertheless, having completed our voyage of discovery around Weizmann's science policy, and in light of his declaration at the opening of the Hebrew University in 1925 asserting the "infant University's" primary tasks as "Patient research in the interests of science, combined with practical service to the people of Palestine and its neighbors," the question remains: To what extent, by dint of the science policy set out by Weizmann and his associates, were his goals realized – in particular, that "our University … win its spurs and build up its reputation by the distinctive value of its contributions to the common stock of knowledge"?[226]

225 Mardor, *Rafael* (above, note 183), pp. 94–96.
226 *Inauguration*, Small Format (above, note 67).

Hedva Ben-Israel

Weizmann and the Hebrew University

Chaim Weizmann devoted his life both to science and to Zionism.[1] At first the two vied for his attention, leading him to write once about how hard it was to serve two gods. But as time went on the two united into a single goal – the establishment of the Hebrew University. For him, launching an institution of higher education in Jerusalem was both a scientific enterprise and a realization of Zionism. In an early letter, he referred to chemistry and the university project as his two loves.[2] The idea had first been broached in the 1880s and lay at the heart of Zionism from the time Hermann Shapira proposed it at the First Zionist Congress. But there can be no doubt that beginning in 1901, when the Fifth Zionist Congress resolved to establish such an institution, no one worked for it as consistently and assiduously as Chaim Weizmann.

In implementing the idea, Weizmann displayed flexibility in his tactics, as evidenced by his relentless efforts to enlist all possible forces,

1 This article is based on three principal sources. For the period up to 1918 I have primarily used the two volumes of Jehuda Reinharz's monumental biography (Reinharz, *CW*, I and II), which I consulted in the Hebrew translations of, respectively, Moshe Rosen and Devorah Barzilai-Yegar (Jerusalem: Zionist Library, 1987; 1996), as well as the three volumes of *The History of the Hebrew University of Jerusalem*, I: Shaul Katz and Michael Heyd (eds.), *Origins and Beginnings*; II: Hagit Lavsky (ed.), *A Period of Consolidation and Growth*; III: eadem (ed.), *Academic Progression in a Period of National Struggle*, Jerusalem: Magnes, 1997; 2005; 2009 (Hebrew). For the period from 1918 onward I have relied principally on my own research for the chapter "Politics on Mount Scopus during the Mandatory Period," *ibid.*, III, pp. 3–86.

2 Weizmann to Vera Khatzman, *CWL*, I, no. 260, pp. 331–332 (July 30, 1902).

both Zionist and non-Zionist. This synthetic approach, mediating between political and practical Zionism, was characteristic of him, but it was especially prominent in his efforts to found the university. In fact, those efforts were replete with apparent shifts and contradictions. For example, at one point he declared that recruiting scholars was more important than the construction of buildings, but at another that it was not intellectuals but workers who would build up the Land of Israel. At yet another juncture he asserted that the establishment of a spiritual and intellectual center of Jewish life in Palestine would lead to the realization of material goals.

Weizmann always emphasized that the university needed to be of the highest scientific quality. However, he also insisted that it would not be an ivory tower but rather a place where a laborer could also complete his studies. He took the same integrating and mediating approach toward both the humanities and the natural sciences. At a time when many thought of the two as separated by an abyss, Weizmann viewed the exact sciences as an organic part of intellectual culture. Like the Zionist teacher he revered, Ahad Ha'am, Weizmann believed that the Jewish people had survived not by virtue of its material assets but thanks to its culture and spiritual values. He maintained this in his youth and reiterated it at the inaugural session of the new State of Israel's Constituent Assembly on February 14, 1949: "We must build a new bridge between science and the human spirit. … [A]bove science there are sublime values, and … only they can cure the ills of humanity – the values of justice and integrity, peace and brotherhood."[3] For Weizmann, the integration of Yavneh and Europe, Judaism and science, humanism and materialism, were among the justifications for Zionism.

This was Weizmann's overarching approach. Historically, his interest in the establishment of a university in Palestine grew out of an earlier debate within the Zionist movement over the role to be played by culture in the Zionist program. Theodor Herzl and Max Nordau did not include Jewish culture in their plans, which were political and social as

3 Weizmann to Ahad Ha'am, *CWL*, I, no. 156, pp. 227–230 (February 5, 1902); *Knesset Proceedings*, I, p. 7.

well as humanistic, in the sense of seeking to improve the character of the individual Jew. But Weizmann and his colleagues in the Democratic Fraction viewed the renewal of Jewish culture as a prime objective. At the final session of the Fifth Zionist Congress in 1901, both Martin Buber and Weizmann argued that Jewish national education and the Hebrew language needed to be essential elements in the Zionist program. The Congress decided on the establishment of a Jewish *Hochschule*, an Institute for Higher Studies.

In the wake of this decision, and following Herzl's declaration that he was sympathetic to the establishment of a university, Weizmann notified Herzl that he would serve as the leader of a loyal opposition. Herzl in turn undertook to promote the university project. He conveyed the idea to the Ottoman sultan's secretary and urged Weizmann to draft a detailed proposal, including a budget, to be presented at a meeting in Turkey. Weizmann, Buber, and Berthold Feiwel proceeded, in 1902, to draw up just such a proposal – a comprehensive plan for a university.[4]

From the start, there was disagreement about the nature of the institution. Was it to be focused primarily on research or on teaching? Was it to be a Zionist or a Jewish university? Was it to serve the Yishuv (the Jewish community in Palestine), the larger world Jewish community, or the Middle East? Should it aspire to the highest standards of science and scholarship or to produce a maximum number of graduates? All these questions were fervently debated. Vladimir Jabotinsky advocated a teaching university; Judah Leib Magnes dreamed of creating a new Judaism and a new Jew; Weizmann wanted a sanctuary that would integrate Zionism, Judaism, and science on a world-class level.

His practical plans underwent many changes over the years. For example, the school of forestry and agriculture recommended in the 1902 paper came into being, as a faculty of agriculture, only in the 1930s.[5] The medical school, to which Weizmann gave first priority in a list he drew up in 1912, was founded only in 1949. Weizmann was concerned

4 Martin Buber, Berthold Feiwel, & Dr. Ch. Weizmann, *Eine jüdische Hochschule*, Berlin: Jüdischer Verlag, 1902.

5 *Ibid.*, p. 32.

about the plight of young Jews who had difficulty entering European universities because of anti-Jewish quotas, and who suffered discrimination even if they were accepted. Consequently, at a meeting of the preparatory committee for the establishment of the university convened in April 1901 in Munich, and on many occasions thereafter, Weizmann proposed the establishment of Jewish institutions of higher education in Russia, Switzerland, and England as well as in Palestine, to offer programs in technological fields, the exact sciences, and Jewish studies. With the Ottoman government opposing its establishment in Palestine, he also changed his mind from time to time about where the university ought to be located. Most of those involved, headed by Menachem Ussishkin, insisted nevertheless that the university could come into being nowhere else – despite their acknowledgment that the Yishuv was numerically and economically weak and that non-Zionist donors to the project opposed Palestine as the seat of the new institution. Weizmann displayed flexibility and was prepared to consider a temporary location outside Palestine, but he always maintained that the university would in the end be built in Jerusalem.

Between 1902 and 1904, when Weizmann's professional standing was not yet on a firm footing but his enthusiasm for the establishment of a Jewish university knew no bounds, he worked feverishly and was exhausted by the competing claims of his commitments to science and to Zionism. In 1902, following an intensive period of lecturing in chemistry, Weizmann set out on a trip through Russia, as he began collecting funds for the university and forming committees to oversee its establishment. The industrialist Israel Brodsky contributed 300 rubles, despite being an opponent of Zionism. Weizmann also carefully studied the histories of other universities. In the same year, the money he collected enabled him to set up a university bureau in his Geneva apartment and to employ a staff to help further the plans.

He worked on three fronts simultaneously. First, he sought to put together a worldwide committee of well-known Jews. Second, he conducted a survey of Jewish professors and students. Third, he tried to launch an academic journal. These initiatives were only partly successful. Among Weizmann's goals was to rescue Jewish students from what he saw as the clutches of socialism and Bundism, which he had long opposed. In this Weizmann held a view prevalent in Germany, Russia, and Britain – that young Jews were particularly susceptible to revolutionary socialism (later, Lord Balfour and Winston Churchill would think the same). Weizmann was attuned to the distress of young Jews who had been alienated from Orthodox Judaism and were turned off by Herzl's seemingly bourgeois brand of Zionism. He hoped that a Jewish cultural and scientific revival would excite the younger generation and eclipse socialist dreams.

The intensive activity of the early years let up to a certain extent in the following period. Weizmann found himself at a crossroads. He tried but failed to organize a summer program in Jewish studies for European students. He thought that such a program would be, as he put it, a wonderful national display. At the same time, he was trying to obtain an academic appointment for himself in England. When the German society Ezra resolved to establish a teachers' college in Jerusalem, Weizmann was a candidate to serve as the science instructor, but that did not come to pass. He visited Palestine for the first time in 1907, at a time when he was wavering between his sincere desire to settle in Palestine and his efforts to obtain British citizenship and an appointment as a professor. When the Technion was founded in Haifa, he hoped to lead its chemistry department, but in the end no such department was set up. Only in 1912, after his economic position had improved and the academic demands on him had lessened, did he devote himself again to the dream of a Hebrew university.

Weizmann's renewed activity was spurred in part by the weakening of the Ottoman regime, the expansion of the Yishuv, the foundation of the Herzliya Hebrew Gymnasium in Tel Aviv, and the support of Baron Edmond de Rothschild for the university. On a personal level, Weizmann was also motivated by his failure to move up the academic ladder in England. In May 1912 he proposed, in a letter to his then admirer Magnes, that all the Yishuv's academic institutions be combined under a single umbrella, along the lines of the Pasteur Institute. He also renewed his attempts to interest prominent Jewish scientists in the establishment of a Jewish university. He was able to put together a committee to work with him, composed of Leo Hermann, Leo Motzkin, Julius Simon, Feiwel, Magnes, and Buber. He also drafted a new working plan, somewhat different from the earlier one, which he presented to Ussishkin and Magnes. The new version called for the establishment of an institute for Oriental and Jewish studies, as well as faculties of political science, law, and medicine. In the hope of obtaining Turkish consent, Weizmann proposed to begin with medicine and law, which he presumed the Ottoman government would appreciate.

But Magnes's dream was of a Jewish spiritual revival, and he thus wanted to begin with archaeology and general humanities, taught from a Jewish point of view. In his view, for example, history and philosophy would not be studied as independent disciplines. Rather, those parts of these fields that contributed to Jewish renewal would be chosen and taught together. (Later, when Magnes became president of the university, he indeed sought to further this idea.) But just as Weizmann began promoting the university once more, public interest began to fade somewhat. Many Zionists, Arthur Ruppin among them, openly advocated turning the movement's efforts toward settlement rather than to founding a university. Others opposed the establishment of a university, claiming that it would only create a redundant or unemployable intellectual proletariat.

Weizmann nevertheless managed to convince the executive committee of the Eleventh Zionist Congress to allow him to address the assembly on the plan. The congress, held in Vienna in 1913, was an important milestone. Twenty delegates from the Yishuv participated, and they demanded higher education in Hebrew. Shmaryahu Levin proclaimed that Jewish education in Palestine should be run by the Zionists. Ussishkin spoke in Hebrew on the establishment of a university, and Weizmann presented his detailed plan, approved in advance by Magnes and Feiwel. The new program, unlike its predecessor, explicitly designated Jerusalem as the location of the university. Weizmann spoke of the need "to create new Jewish values and to integrate our great traditions with the values of the new era."[6] The Orthodox delegates expressed their

6 *Stenographisches Protokoll der Verhandlungen des. XI. Zionisten-Kongresses*

concern about the application of scholarly methods to the study of Jewish sources, but they supported the training of doctors and scientists. Weizmann was appointed to chair the University Committee, and in his speech he famously asserted that the institution would be a spiritual dreadnought, one that would achieve more than other nations do with armies and fleets.[7] The idea that science would be the guarantee of future Jewish life, just as the culture of the Jews had preserved them in the past, is one that Weizmann reiterated frequently.

Weizmann and Magnes were close at this time. Both believed in the creation of a new Jew, "with his head held high and exalted by creative powers," the product of an intellectual center that would fashion a new, scientific Jewish culture. The university would attract immigrants to Palestine, take in students excluded from European universities, enhance the reputation of the Jews in the world, train teachers for the Diaspora, and provide trained professionals to the Turkish empire and the Middle East.[8] The difference between the two men was one of emphasis and preference. Magnes for the most part stressed the renewal of Jewish culture, while Weizmann also emphasized pure science. The program Weizmann presented to the Zionist Congress in 1913 noted that the scientific infrastructure for the establishment of a medical school was already present in Palestine, and that Jewish studies would, of course, be studied in Hebrew. But it was still not clear what the language of instruction would be in mathematics and the sciences. The language problem would soon erupt in the Yishuv over the question of whether Hebrew should be used in instruction at the Technion and its associated high school. The demand for Hebrew, eventually accepted, led to a fissure between the Zionists and the non-Zionists and also distanced Zionism from German influence.

Despite these disagreements, Weizmann stuck to the plan to begin with a medical institute. To that end, he enlisted the support of Baron

in Wien, vom 2. bis 9. September 1913, Berlin–Leipzig: Jüdischer Verlag, 1914, p. 302.

7 The dreadnought was an innovative British warship, introduced in 1906.

8 *Stenographisches Protokoll* (above, note 6), p. 303.

Rothschild. They met in January 1914, and Rothschild made a large donation, but he demanded the establishment of research institutes and not a teaching institution. Weizmann, in his contacts with the baron, used tactics much like those previously used by Herzl himself: He showed himself willing to make personal appeals to those with power and money, and also to compromise with those whose support he sought. He agreed with Rothschild on the importance of research institutes, but, along with Ahad Ha'am and Magnes, he consented to postpone teaching only for the time being. Weizmann finally succeeded in persuading Rothschild to accept that in the future the university would also engage in teaching; if it did not, he pointed out, young European Jews would have no choice but to convert to Christianity in order to obtain a higher education. He was proud of his success with the baron, but he was criticized by Jabotinsky and others, who demanded a teaching university immediately, even if this meant losing Rothschild's support. Weizmann was taken aback by the ferocity of the opposition and argued that both Zionism and the establishment of a Hebrew university in Jerusalem had to be seen as evolutionary historical processes rather than immediate and urgent responses to antisemitism. He maintained that the establishment of research institutes was part of a process that would lead to the creation of a center of science and might even bring the institution Nobel prizes. The debate continued as Ruppin conducted negotiations to buy land for the university on Mt. Scopus. Then the war broke out and put an end to all this activity.

Even before the war, Zionists were divided as a result of their patriotic feelings toward their countries of residence. Baron Rothschild, for example, was a loyal French citizen, while Isaac Strauss, an industrialist, a friend of Weizmann's, and an important donor, was a steadfast German. During the war the German government sent him to the United States to persuade that country's Jews to oppose American entry into the war. The language wars in the Yishuv also frustrated Weizmann's attempts to bridge between German Zionists, who wanted German to be the language of instruction, and other western Zionists, who advocated Hebrew. These fractures worsened after the war broke out, with Jews serving in the armies of each side. Until that point, Weizmann's position on the use of Hebrew had not been extreme, for both tactical reasons (he needed the support of German Jews, such as Paul Ehrlich) and practical ones (he was not sure that Hebrew was adequate for instruction in science). But he believed deep down that Hebrew would win out in the end. On the eve of the war, he sensed that his life's mission was close to being realized, but he certainly did not imagine that the war itself would catalyze the process and enhance his leadership. He anticipated neither the British conquest of Palestine nor the opportunity that fell into his and Zionism's lap thanks to his scientific contribution to the British war effort. The latter gave him cachet and helped him achieve the Balfour Declaration, while also demonstrating the close tie between science and Zionism, which had always been Weizmann's dream.

But those steps forward came only toward the end of the war. At first, the war worsened the internal fissures in the Zionist movement, including those between Weizmann and his supporters. One of these was Magnes, with whom he had worked harmoniously up to this point. In his later reflections on this period, David Ben-Gurion put considerable stress on Weizmann's anglophilia and his view that English support was vital for the Zionist cause.[9] Magnes, by contrast, was a fervent pacifist whose advocacy of neutrality in the U.S. had cost him his standing in the Jewish community. He criticized both Weizmann and Jabotinsky for their overdependence on Britain, which he condemned as an imperialist power. Even before the war, Magnes had accused the two others that their fealty to Britain was exacerbating Ottoman hostility toward the Yishuv. During the war he opposed the establishment of the Jewish Legion. Weizmann, for his part, did not oppose fighting alongside the British and went so far as to sever his connections with Zionists in the countries fighting against the Allies. Following the Balfour Declaration, the British conquest of southern Palestine, and an audience with the king, Weizmann set off for Palestine with the Zionist Commission, in part to lay the cornerstone for the university.

9 David Ben-Gurion, "Preface," in Meyer W. Weisgal & Joel Carmichael (eds.), *Chaim Weizmann: A Biography by Several Hands*, New York: Atheneum, 1963, p. 4.

Weizmann's loyalty did not prevent the British army from constraining him. He did not easily obtain British sponsorship for the cornerstone-laying ceremony in July 1918, and when it was granted, General Clayton forbade Jewish soldiers in uniform (including James de Rothschild) to lay one of the stones "in the name of the Hebrew army." In his speech at the ceremony, Weizmann stressed the centrality of academic achievement in Jewish rebirth and promised that the university would be open to all. He concluded with a picturesque fantasy that the Jewish soul hovering between heaven and earth would finally find a resting place and a home in this sanctuary of science and scholarship.[10] In the same spirit, in 1920, speaking of his expectations of the university's scientific work, he offered his vision of Zionism as spiritual renewal rather than merely a territorial and political ingathering. He wrote that the Russian Jewish academic, caught up in his own personal development, fostered a sterile and egocentric intellect and lived in what was for all intents and purposes a spiritual ghetto, alien to the world around him. The Hebrew University in Jerusalem, by contrast, would produce Jewish scientists with roots deep in their own social and cultural soil.[11]

The years from 1918 to 1925 were critical ones for the university. During this period, Weizmann served as president both of the university's board and of the Zionist movement. His efforts to recruit scientists for the establishment of the medical faculty, and then for the chemistry faculty, led to the institution's first academic appointment, of Professor Andor Fodor, a biochemist. In 1920, a committee headed by James Rothschild and Otto Warburg drew up plans for three natural science institutes – physics, chemistry, and microbiology – as well as for an institute of Jewish studies and humanities. Weizmann, assisted by Albert Einstein and Alfred Zimmern, a British classicist and historian, focused on recruiting the best Jewish scholars for the university. In 1924 he

10 For the original version of the speech, see B.Z. Kedar, "Laying the Foundation Stones of the Hebrew University of Jerusalem, July 24, 1918," in Katz and Heyd, *History of the Hebrew University* (above, note 1), pp. 110–119.

11 Weizmann to Abraham Tulin, *CWL*, X, no. 16, pp. 27–29 (August 22, 1920).

also tried unsuccessfully, together with Einstein and Ahad Ha'am's son Shlomo Ginzberg, to raise funds in the United States.

Among the opponents of the project was the Vatican, which feared that a Hebrew university would encroach on the Catholic educational institutions in Palestine. Weizmann saw another potential source of competition in a program put forth by the Red Cross to improve the country's sanitation, which he countered by initiating a Zionist research project in the same field. In all these cases, he was guided by his confidence that scientific achievements would produce the conditions for building the Jewish homeland. Yet he was criticized for his single-minded focus on the university. The years following the Balfour Declaration were critical ones for achieving Zionist objectives, and Nordau and Jabotinsky attacked him for not pursuing massive Jewish immigration, which they considered vital to guaranteeing that Palestine would become a Jewish territory.

During the university's early years, contention grew over control of it, in particular between Weizmann and Magnes, who as chancellor was responsible for administration. In 1923, Magnes used money he had raised himself to found the Institute of Jewish Studies. He became the unofficial representative of donors, both Zionist and non-Zionist, from the United States. Weizmann, who viewed the university as his life's greatest project, was compelled, from that time onward, to cope with Magnes's consistent policy of seeking to make the university ideologically, academically, and financially independent of the Zionist movement. The two men, who had long cooperated with each other, now became fierce rivals. This could be seen even at the opening ceremony in 1925. Magnes went so far as to criticize the ceremony, which garnered much public enthusiasm in the Yishuv. He found fault with what he saw as the political character of the ceremony, manifested in particular by the invitation of Lord Balfour to speak. As far as Magnes was concerned, this constituted a provocation against the Arabs and a capitulation to the forces of imperialism.[12]

Magnes was a man of moral and intellectual integrity, supremely

12 See Arthur A. Goren (ed.), *Dissenter in Zion: From the Writings of Judah L.*

self-assured and ideologically unswerving. He constantly reiterated that he and not Weizmann was the right man to head the university, taking his critique of his erstwhile colleague to such an extreme as to claim that Weizmann's involvement in university affairs, as a leader of the Zionist movement, constituted illegitimate state (his word!) interference in matters of education and conscience. Weizmann was no less outspoken in his censure of Magnes. In 1929, the fateful year in which bloody Arab attacks on Jews left about 140 men, women, and children dead and 340 wounded, Weizmann wrote angrily about the hypocritical Magneses who came out in defense of Arab terrorists and went so far as to plead for clemency for those who had been sentenced to death. Weizmann maintained that the university had to be closely tied to Zionism. True, his profound belief in the importance of a Jewish spiritual and intellectual center of the type advocated by his revered mentor, Ahad Ha'am, enabled him to maintain useful ties with non-Zionists, but he never gave in to demands to sever the university from the Zionist Organization. The debate over the financial support provided to the university by the Zionist institutions, in return for influence over it, continued for many more years and came up for discussion at the Fourteenth Zionist Congress in 1925. In practical terms, however, the Zionist contribution to the university budget was scaled back considerably, and the university gained considerable independence from the movement.

Weizmann and Magnes differed on both politics and academic policy. Magnes took advantage of his speeches at the opening and closing ceremonics of the academic years to openly advocate binationalism (1929) and to lambast the Yishuv's self-defense organizations. In a fiercely radical speech in 1933, spurred perhaps by the Nazi rise to power, Magnes renounced cooperation with the Zionist leadership and maintained that the university should not be sullied by chauvinism and militarism. Another notable political clash between him and Weizmann took place at the fifth session of the Jewish Agency Council in Zurich in 1937, in connection with the Peel Commission's proposal to partition Palestine.

Magnes, Cambridge, MA–London: Harvard University Press, 1982, nos. 48–49, pp. 231–235.

Magnes attacked the proposal, and Weizmann responded angrily. "Are all of us here, excepting Magnes, land thieves?" he asked, warning ironically against those good Jews who were telling the world that only the evil Jews from eastern Europe wanted a Jewish state. They had done so in 1917, he declared, just as they were doing in 1937. The good Jews had remained good, while the "bad" Jews who advocated a Jewish state had drained swamps, built schools, and founded the university.[13] But despite these disagreements and the disputes that broke out between them from time to time, neither Magnes nor Weizmann despaired of persuading the other that he was right. In the 1930s, Magnes infuriated the Jewish Agency leadership by meeting secretly with Arab leaders and mediators in an effort to seek an accommodation between the two peoples. He nevertheless did all he could to gain Weizmann's support and sent him detailed proposals for achieving peace with the Arabs and founding a binational state.

Most of the criticism of Magnes's politics came from the Zionist institutions and the Yishuv leadership. By contrast, critiques of his academic policy, his appointments, and his authoritarian style were voiced by the university's most loyal supporters around the world, most prominently Einstein. Einstein was, in fact, an admirer of Magnes's politics, as he wrote on the occasion of the latter's sixtieth birthday in 1937. But he was a consistent, strident, and acute critic of the way the university was being run. He said time and again that Magnes was not qualified to appoint scientists, and he voiced his concern that the institution, under Magnes, would become a farmers' university or at best an American college. So disappointed was Einstein that he resigned in 1928 from the leadership of the university's academic committee, leading Weizmann to assume this position in addition to his other posts.[14]

Despite Weizmann's attempts to act with moderation, relations

13 The Twentieth Zionist Congress and the Fifth Session of the Jewish Agency Council, Zurich, 26 Av–14 Elul 5697, Part II, pp. 58–74; Goren, *Dissenter in Zion* (above, note 12), pp. 329–334.

14 [For a detailed discussion of the relations between Weizmann and Einstein, see Issachar Unna's article in this volume. – Ed.]

between him and Magnes worsened, both politically and personally. In 1931, when Weizmann was ousted from the presidency of the Zionist Organization, he indicated his willingness to settle in Palestine and work at the university, but his inquiry received no response. In 1933, the severe criticism leveled by Einstein and others at Magnes's academic administration hurt the university's fundraising efforts just as it was trying to take in refugee scholars from Nazi Germany. Weizmann and Einstein both advocated an inquiry into the university's governance. This resulted in the Hartog commission, which found fault with the administration. In the wake of the commission's report, the university's charter was amended, and control of its affairs was placed, for all intents and purposes, in the hands of the academic staff and its elected representatives. The post of chancellor was eliminated, thus ending Magnes's autocracy. He was made the university's president, a post almost devoid of authority.

The upshot was that both rivals found themselves bereft of any power over the university, which became an autonomous institution. Einstein's fear that one dictator would simply be replaced by another turned out to be unfounded. It was during these reforms that Weizmann founded the Sieff Institute in Rehovot. Together, these two developments minimized the areas of friction between him and Magnes. In fact, it may well have been his conflict with Magnes that led Weizmann to persuade the Sieff family to establish a scientific institute in Rehovot in memory of their son, separate from the Hebrew University. The Institute, which would become the kernel of the Weizmann Institute, became Weizmann's refuge for his own scientific work. However, the new research institute ran into some of the same problems encountered by the university in Jerusalem, including the issue of whether it should engage in instruction. Another point of similarity was that, just as Magnes, seeing Weizmann's interference in university affairs as political, had battled against it, so Weizmann fought the Israeli government's attempts to intervene in the affairs of the institute in Rehovot.[15] He was also disappointed there, as

15 [On this, see Benny Morris's article in this volume. – Ed.]

he had been at the Hebrew University, that major scientists who had fled Nazi Germany did not join the institute.

On November 2, 1949, at the ceremony inaugurating the Weizmann Institute, Weizmann was lauded both by Selig Brodetsky, professor of mathematics and the new president of the Hebrew University, and by Prime Minister David Ben-Gurion, who declared that Weizmann, now president of Israel, wore the crowns both of wisdom and of national leadership. A few days later in London, at a celebration of Weizmann's seventy-fifth birthday, the keynote speaker, General Jan Smuts, extolled Weizmann and declared that, at a time when all of humanity was in dire straits, some consolation could be found in the events in Israel, with its spiritual and scientific response to distress. He also said that Israel's and Britain's victories in their wars of survival were among the great achievements of the era.[16]

Even before, during and after World War II, Weizmann and Magnes grew closer. The university was already self-governing, and despite Magnes's personal opinions, it fully identified with Zionism and the Yishuv. Both men proudly and fondly remembered their common labors and earlier friendship. They also worked together to rescue libraries and cultural artifacts that the Nazis had stolen from Jews. There was even some political convergence, with both of them expressing repugnance at the use of violence by the Jewish underground forces. In 1944, when the two met during a visit by Weizmann to Jerusalem, they both voiced disgust at the murder of Lord Moyne and agreed that Churchill could not be expected to support the establishment of a Jewish army if its weapons might be aimed at the British. During the visit, Weizmann also listened patiently to Magnes holding forth on peace with the Arabs, and to his claim that nationalist education in Yishuv schools was fostering violence.

One dramatic incident illuminates the relations between Magnes and Weizmann. In 1946, they went together to visit Henrietta Szold, the founder and leader of Hadassah, the Zionist women's organization, who was in hospital. Each held one of her hands, and she told them how

16 Weisgal & Carmichael, *Chaim Weizmann* (above, note 9), p. 352.

happy she was to see them together. When they left her room, they promised each other not to quarrel any more. In 1947 the university awarded both of them honorary doctorates. Magnes never abandoned his opposition to a Jewish state, and in 1948, during the fighting, he worked as hard as he could in Jerusalem and in Washington to prevent its establishment. Nevertheless, on the day the state was declared, he sent Weizmann his congratulations.[17] After the United Nations decision to internationalize Jerusalem, when the university's faculty was determined to keep the university in the city, Weizmann considered transferring it to Tel Aviv, so that it would be located within the Jewish state.[18] But Jerusalem did not in the end become an international city, and the university remained there. Weizmann became Israel's president, and Magnes died soon after.

Both Judah Magnes and Chaim Weizmann left permanent imprints on the Hebrew University – Weizmann in his devotion to a high level of scholarship and the revival of Jewish culture, and Magnes in his moral commitment to self-governance and freedom of inquiry. Together they fashioned a pluralistic research institution of the highest order, as it remains today.

17 Goren, *Dissenter in Zion* (above, note 11), p. 55.

18 Weizmann to Werner Senator, *CWL*, XXIII, no. 232, p. 185 (July 25, 1948); Weizmann to Abba Eban, *ibid.*, no. 235, pp. 188–189 (July 27, 1948); Weizmann to Senator, *ibid.*, no. 240, pp. 194–195 (August 1, 1948).

Issachar Unna

Wealth or Genius?

The Struggle of Weizmann and Einstein for Academic Excellence at the Hebrew University

Sixty years ago, Rabbi Samuel Belkin, president of Yeshiva University in New York, invited Albert Einstein for a visit. When the esteemed guest arrived from Princeton, Belkin took him on a tour of the building designated for the university's new medical school. "We plan to call it the Albert Einstein College of Medicine," Belkin told his visitor. "Why Einstein?" the guest asked. "I don't know a thing about medicine." Belkin replied: "If you don't agree, we'll name the college after David Abramson. "Who is Abramson?" Einstein asked. Belkin answered: "You see!"

Belkin was smart. He knew that "Albert Einstein" was a valuable brand. On his seventy-fourth birthday, Einstein agreed to allow his name to be used. Today, everyone in New York and in the medical field knows about Einstein Medical, the top-flight medical school in the Bronx.

Chaim Weizmann was the first to appreciate the value of the Einstein brand. Einstein initially made newspaper headlines in 1919, in England in particular, when the results of a British solar eclipse expedition to the tropics published measurements that confirmed one of the predictions of his General Theory of Relativity. Weizmann immediately realized that Einstein had to be enlisted in the Zionist cause. He also grasped that the best way to do so would be to interest him in the idea of establishing the Hebrew University, a university for the Jewish people, in Jerusalem.

But Weizmann could hardly have understood what the cost of his success would be. Unlike Belkin and his medical school, Weizmann did not simply get permission to use Einstein's name. Einstein put his heart and soul into the university project. Weizmann had enlisted a fighter for his cause of making academic excellence the new institution's guiding principle. But that fighter sometimes fought alongside him and sometimes against him. For Einstein, responsible academic administration aimed at ensuring excellence was an inviolable condition for his cooperation. Weizmann unquestionably shared that ideal, but faced constraints, in particular regarding the demands and interests of the project's American financial backers. Weizmann spent years navigating between the Scylla of American capital and the Charybdis of the famous physicist whose imprint Weizmann was not willing to lose.

In 1921, Weizmann asked Einstein to join him on an American fundraising tour.[1] Einstein agreed. The two men presumably got to know each other well during the eleven-day sea voyage from Europe to New York and their subsequent joint appearances before potential donors.[2]

The next important step in relations between the two men came in the autumn of 1923. "I am delighted to hear from [Leonard] Ornstein [a Dutch physicist involved in the university project, who later headed its physics department]," Weizmann wrote to Einstein on October 24, "that you are prepared to accept the presidency of the university Board of Governors. The obvious conclusion is that you will join the Jewish Agency leadership." Weizmann laid out in his letter the other persons slated to make up the new body, which was meant to unite Zionist and non-Zionist forces to promote the welfare of the Yishuv, the Jewish community in Palestine.[3]

Three days later, Einstein wrote that he had been delighted to receive a personal letter from Weizmann. He conditioned his acceptance of the

1 Weizmann to Einstein, AEA, 70-956 (February 23, 1919); Diana K. Buchwald et al. (eds.), *The Collected Papers of Albert Einstein* (*CPAE*), XII, Princeton: Princeton University Press, 2009, doc. 63.

2 [For a detailed account of Einstein's trip and the objections it aroused, see Shulamit Volkov's article in this volume. – Ed.]

3 Weizmann to Einstein, AEA, 33-365 (October 24, 1923); *CPAE*, XIV, 2015, doc. 131. Many of the quoted documents can also be found in the Weizmann Archive in Rehovot and were published in Weizmann's collected letters.

offer on not being required to travel. He agreed to the use of his name and said that he would attend local meetings when he was in Berlin. He wrote that his trip to Palestine the previous February had made an enormous impression on him, and expressed his full confidence that the Zionist enterprise would succeed in the end, even if it was at the cost of sweat, sacrifice, and disappointment.[4]

Weizmann wrote to Einstein a month later that he had been pleased to receive his consent, and that the conditions Einstein had laid down would be fully respected. He explained how important their work was, especially in light of European Jewry's desperate plight (this in 1923!). He informed Einstein that a scientific institute would soon open its doors on Mt. Scopus and hoped that Einstein would attend the inaugural ceremony that coming May.[5] In January 1924, Einstein received a detailed letter from David Eder, a leading British Zionist, reporting on progress in putting together the Board of Governors. Among its members would be Ahad Ha'am, Weizmann, Baron Edmond James de Rothschild, Sir Alfred Mond, and leading American scholars and philanthropists whom he did not mention by name. A university senate, the Scientific Council, would also be established, to serve primarily as an advisory body on academic affairs. The letter also reported the progress of the construction of the institutes of microbiology and biochemistry.[6] Two months later Einstein turned down a request from Weizmann and the American Jewish Physicians Committee to make another trip to the United States. He wrote that he was happy to hear that they had succeeded in bringing what he referred to as "the leading Jewish goyim" of America back into the fold.[7] In June 1924, Einstein again rebuffed Weizmann's repeated pleas that he attend a meeting, mainly regarding university affairs, scheduled for July 13 in London. He believed at the time that he and Judah Leib Magnes, the university's designated chancellor, were in agreement about how the university should be run. The principal

4 Einstein to Weizmann, AEA, 33-366 (October 27, 1923); *CPAE*, XIV, doc. 133.

5 Weizmann to Einstein, AEA, 33-367 (November 19, 1923); *CPAE*, XIV, doc. 212c.

6 Eder to Einstein, AEA 26-869 (January 9, 1924); *CPAE*, XIV, doc. 194.

7 Einstein to Weizmann, AEA, 33-371 (February 29, 1924); *CPAE*, XIV, doc. 219.

reason he could not take part in the conference was his previous commitment to the International Committee on Intellectual Cooperation of the League of Nations, which was convening in Geneva.[8]

In a long letter written on December 12, 1924, Weizmann reported to Einstein on the impressive progress and development of the Jewish settlement effort. He made special note of the advances toward the establishment of the university and of the Technion in Haifa, both of which were to open soon. He asked for Einstein's active assistance in establishing ties and cooperation between these new Yishuv institutions and important universities, research institutes, and libraries in Europe.[9]

Einstein sowed the first seeds of the approaching crisis in a letter he wrote to Weizmann just before setting out for a trip to South America. He laid out in detail how he thought the university should be run. He stressed that the administration had to be made up of world-renowned (that is, European) scholars. This was the only way to gain and ensure the respect of the Diaspora intelligentsia for the new university. He stressed – so strongly that he skirted offense – that, on the basis of his impressions of the situation in Palestine during his visit, the administration should not be centered there.[10] Einstein repeated all this more explicitly, and more belligerently, in a letter he sent directly to Magnes prior to the second meeting of the Board of Governors. "The intellectual competence of the scholars living in Palestine is insufficient to address the determination of the budget and the appointments of scientists," he asserted.[11]

Einstein celebrated the opening of the university on April 1, 1925, with the Jewish community of Buenos Aires, at a large event, attended by thousands, at the Teatro Coliseo.[12] He also wrote an article on the university's mission, published in English in Palestine and in a German

8 Einstein to Weizmann, AEA, 33-374 (July 16, 1924); Einstein to Weizmann, AEA, 33-375 (July 19, 1924); *CPAE*, XIV, docs. 286, 288.

9 Weizmann to Einstein, AEA, 33-376 (December 12, 1924); *CPAE*, XIV, doc. 394.

10 Einstein to Weizmann, AEA, 67-633 (March 2, 1925). See also AEA, 71-128 (the letter is preserved in the archive in two copies); *CPAE*, XIV, doc. 451.

11 Einstein to Magnes, AEA, 36-895 (September 9, 1925).

12 See Einstein's travel journal, AEA, 29-123 (April 6, 1925); *CPAE*, CIV, doc. 455.

Zionist newspaper.[13] Einstein also missed the Board of Governors meeting that convened three weeks later in Tel Aviv. He must certainly not have been pleased when he read the record of that meeting, which included the appointment of Magnes as chancellor, signalling that the Americans were taking control of the university administration.

At the beginning of July 1925, Einstein received a long, pessimistic, and confidential missive from Weizmann, who reported that he had resolved to resign from the presidency of the Zionist Organization and from the university's Board of Governors. Because of their friendship, he wrote, he thought it only proper for him to inform Einstein of his intentions before making them public, so that he would not learn the news from the press.[14] Einstein responded immediately. "Your letter is written tactfully, but the depressing tone was a painful surprise for me," he wrote. "Your work seems so brilliant to me, I'd almost say detached from the world's gravity." Later in the letter he supported Weizmann's decision to resign, but only as a tactical move aimed at removing the obstacles that his rivals had placed in his path. He did not under any circumstances consent, however, to Weizmann's resignation from the university administration. "I of course suspect that neither are you prepared to accept the moves by the rich American donors to get rid of Europe," he declared. For precisely that reason he implored Weizmann not to resign; his influence was vital.[15]

The beginning of the real rift, with Einstein on one side and Magnes and the university administration on the other, can be dated to the Board of Governors' second meeting, in Munich on September 23, 1925, with Einstein in attendance for the first time.[16] Weizmann was in the United States. Einstein emerged from the meeting in a state of shock at the way Magnes had conducted it, but his anger really erupted when he received the minutes, which Magnes had recorded and written up. The record

13 *The New Palestine*, March 27, 1925; *Jüdische Rundschau*, April 3, 1925.

14 Weizmann to Einstein, AEA, 33-377 (June 30, 1925).

15 Einstein to Weizmann, AEA, 33-378 (July 2, 1925).

16 The meeting was held in the home of Rachel and Eli Straus, my father's aunt and uncle (I.U.).

differed notably from the decisions that Einstein understood to have been made. He believed that Magnes had falsified the record in keeping with his own aspirations. It was the final straw.[17] Weizmann brought Einstein's position to Magnes's attention, adding that he agreed with it. The letter clearly detailed the discrepancies between the decisions made at the meeting, according to Einstein's report, and the written record. The two, as Weizmann put it, were "considerably at variance."[18]

Einstein, in a letter to Weizmann dated April 9, 1926, said he was happy that Weizmann was resigning from the Zionist Executive with the intention of devoting himself to the university. He asked Weizmann to consult with him on the ramifications of the influence that the American funders were wielding over administrative policy. "We should discuss this before the coming meeting of the Board of Governors, so that we are on the same wavelength," he urged.[19] That same day, Weizmann penned an exhaustive letter to Einstein in which he went into great detail about the university's sorry academic state. He did not hold back his criticism of Magnes, who, he asserted, was acting like an amateur in his running of the university's Jewish studies. "Unfortunately, I still see no possibility of replacing M.... We must break Magnes's autocracy," he proclaimed.[20] On July 6, Einstein wrote to Weizmann: "I am adamant in my opinion – cooperation between me and Magnes is impossible. So long as he manages the work in Palestine, I will not be able to be actively involved in university affairs." He requested that a letter stating this be circulated among the participants in the upcoming Board of Governors meeting in London.[21]

The crisis worsened in the years that followed. Weizmann tried to persuade Einstein – they were still friends at the time – that he would see to it that the situation changed, but nothing actually happened.

17 The minutes can be found in AEA, 89-854 (Hebrew version), and AEA, 91-408 (German version).

18 Weizmann to Magnes, AEA, 89-112 (December 15, 1925).

19 Einstein to Weizmann, AEA, 33-382 (April 9, 1926).

20 Weizmann to Einstein, AEA, 33-383 (April 9, 1926).

21 Einstein to Weizmann, AEA, 37-742 (July 6, 1926).

Einstein, who had resigned from all his university positions, continued to be pained at the way the university was being run, writing Weizmann personal letters containing his thoughts and proposals for reforming the institution. He luridly recounted how Magnes was destroying the university, basing himself on information he received from his sources. Again and again he wrote that competent academic administration was vital to save the institution.[22]

On June 6, 1928, Weizmann reported to Einstein about the London session of the Board of Governors, which had just adjourned. Weizmann (and thus Einstein, behind the scenes) lost the debate over the proper relationship between teaching and research. The two thought that teaching should be put off for several years so that the university could devote itself solely to research. "But most of the members, and especially the scholars (Landau, for example[23]) maintained that teaching should commence immediately," Weizmann recounted. The Board of Governors decided to appoint an advisory committee on the subject, headed, so the governors hoped, by Einstein.[24] Einstein responded quickly to the governors' request. He informed them that, given the reports he had received from Weizmann, he had decided to completely sever, de facto, his involvement in university affairs.[25] Less than a week later, Einstein informed Weizmann that he saw no reason for Weizmann and Selig Brodetsky (a mathematician and senior British Zionist) to visit him and said bluntly that they should not come. He cited his health but added that he saw no way that he could continue to bear responsibility for what was going on at the university. "I insist that you strike my name from the Board of Governors list," he wrote.[26] The break between the two men seemed final.

22 Einstein to Weizmann, AEA, 33-386 (January 8, 1928); Einstein to Weizmann, AEA, 33-391 (May 29, 1928).

23 Edmund Landau, an acclaimed mathematician and head of the university's Institute of Mathematics.

24 Weizmann to Einstein, AEA, 67-599 (June 6, 1928).

25 Einstein to Weizmann, AEA, 33-395 June 14, 1928).

26 Einstein to Weizmann, AEA, 33-397 (June 20, 1928).

Einstein lashed out at Weizmann in a letter he sent to Fritz Haber on August 9, 1933, in response to Haber's entreaty that he reconcile with his erstwhile friend. Only cooperation between the two could pave the way for him and Richard Willstätter to join the university in Palestine, Haber told him.[27] Einstein minced no words:

> Who would have imagined that my dear Haber would appear as advocate of the Jewish cause and immigration to Palestine! Really, the old fox W. chose an excellent solicitor. But the information he gave you is bad. I thus have to add and correct. The worst thing is that the wonderful Felix Warburg has succeeded, thanks to his financial power, in getting the romantic and incompetent Magnes, a failed [*verkrachter*] American rabbi, appointed director of the institution. His amateur political activity had become an annoyance for his influential family in America, so they sought a respectable exit for him to an exotic world. This selfish and weak man has surrounded himself with other morally inferior people who allow no honest person to advance. The worst of them is a man named Kligler, also an American import, a schemer lacking all scientific substance. They have all caused the atmosphere to be poisoned totally. And they have kept the institution at the lowest level. Five years ago, when I considered the situation intolerable, I demanded their dismissal. Weizmann promised he would resign together with me from the administration if our demands were not met. I did so, but Weizmann broke his promise. I have had no connection with the institution since then .... Weizmann tried again to take advantage of my good name to advertise that pigpen [*Saubude*] for fundraising purposes. I was thus compelled to make a public statement that I did not want any connection with the matter as long as a thorough cleaning [*gründliche Sanierung*] was not accomplished. Weizmann compelled me to take this step when he used my name improperly.

27 Haber to Einstein, AEA, 12-385 (undated). [For more on Weizmann's relations with Willstätter and Haber, see Shulamit Volkov's article in this volume. – Ed.]

Einstein went on to explain that Weizmann would also be delighted to be rid of this incompetent management, but that political and personal constraints were preventing him from doing so. "Weizmann is an intelligent and charming man, but mendacious [*verlogener*]," he charged. He rebuffed the other side's suggestion that his own actions were personally motivated and declared that he was not willing to have any contact with Weizmann or any of his associates until a thorough clean-up of the university had been accomplished. "Please tell Weizmann," he wrote to Haber, "that my trust in him will return only when he proves by his actions that he is worthy of it." He also advised Haber to keep his distance from the Hebrew University as long as no fundamental change was made in the way it was run. On top of that, he expressed his hope that Haber would not return to Germany, where scientists, with only a handful of exceptions, such as Max Planck and Max von Laue, genuflect to criminals.[28]

In the end, Einstein's rage had the intended effect. The Hartog commission was appointed and charged with investigating the university's administration. It submitted its report in 1934, and the Hebrew University's governance was reformed in light of its recommendations. Einstein agreed to rejoin the Board of Governors in 1935. Fifteen years later, on the occasion of the university's twenty-fifth anniversary, he sent his warm congratulations.

I conclude with an exchange of letters between Einstein and Weizmann from the autumn of 1949. On October 20, Einstein wrote: "I send my friendly greetings to you and the members of the Weizmann Institute on the occasion of the official opening. In light of what is known about the institute, we have grounds for hoping that it will make a significant contribution to Israel's economic and spiritual prosperity."[29] Weizmann replied: "I was touched by your consideration in sending your congratulations on the great occasion of the inauguration of the Institute of

28 Einstein to Haber, AEA, 12-387 (August 9, 1933).

29 Einstein to Weizmann, AEA, 71-291 (October 20, 1949). [See Illustration 29 in this volume. – Ed.]

Physics and Physical Chemistry. It hardly needs to be said with what reverence they were received. It was a wonderful event. My only regret is that you could not participate."[30]

These two great men were once again at peace with each other.

30 Weizmann to Einstein, AEA, 71-293 (November 25, 1949).

Ruth Arnon

The Weizmann Institute of Science

A Living Memorial to the Founder of Scientific Research in Israel

It was 1933, a fateful year. Hitler came to power and a wave of antisemitism swept Germany, threatening the country's Jewish scientists. Chaim Weizmann, then president of the British Zionist Federation and president of the Central Bureau for the Resettlement of German Jews, was taking a walk in Hyde Park together with a friend from Manchester, Israel Sieff. Sieff had recently and tragically lost his eighteen-year-old son, Daniel. He asked Weizmann's opinion about an appropriate way to honor his son's memory. Weizmann proposed the establishment of a scientific institution in Palestine, one that could take in Jewish scientists fleeing Germany. The appropriate location, he suggested, would be Rehovot, then a small village surrounded by citrus groves. An Agricultural Experimental Station was already located there, and Weizmann believed that it and the new institution could usefully be integrated. He was confident that biochemical research could benefit agriculture in the country, as well as the citrus industry that had developed in the Rehovot area.

Sieff agreed, and in 1934 the Daniel Sieff Institute was founded, with Weizmann at its helm. At the same time, Weizmann purchased for himself a nearby plot of land and built a spacious home, designed by the famous architect Erich Mendelsohn. Weizmann and his wife Vera lived there for the rest of their lives, and both are buried nearby. The house also served as the official residence of the president of Israel while Weizmann served in that post. Today, with its original furnishings, it serves as the Weizmann Museum.

The Sieff Institute's opening ceremony was an impressive one. The guest of honor was the chemist and Nobel Laureate Richard Willstätter, and other honored guests included Sir Arthur Wauchope, the British high commissioner for Palestine, as well as senior Zionist figures, including Nahum Sokolow, president of the World Zionist Congress, and the poet Hayim Nahman Bialik.

Weizmann's right-hand-man and organizational deputy at the Sieff Institute was Meir Weisgal, who would later be the moving force behind the establishment of the Weizmann Institute of Science. Once the Sieff Institute was founded, Weizmann proceeded to recruit its faculty. He planned to bring in two world-famous scientists, Willstätter and Fritz Haber, but the former decided to remain in Germany and the second died in 1934, while contemplating a trip to Palestine. But Weizmann did not despair. He managed to bring in a staff of about ten in addition to himself. The most senior figure among them was Ernst David Bergmann, who came from Berlin and later became scientific director of the Institute. He was followed by his brother Felix Bergmann, a chemist and physician, who later became a member of the pharmacology faculty at the Hebrew University. Erich Herlinger and Bruno Rosenfeld, both chemists from Munich, also joined. Other recruits were chemists Yehuda Hirshberg from Brussels, Leon Haskelberg from Paris, Esther Hellinger from London, and Hannah Weizmann, Chaim's sister, who came to the Institute from Zurich. It was an impressive faculty for a brand new institute. Each scientist received a laboratory. (Weizmann's own laboratory has been preserved in its original form and is open to the public.) The major research fields in the Sieff Institute were organic chemistry and biochemistry, since Weizmann saw research in these disciplines as combining science and Zionism. "I am certain that science will bring peace to this land and restore its youth, and that it will also create a source of a new spiritual and material life," Weizmann proclaimed in his speech at the Sieff Institute's opening ceremony. He stressed that "in speaking of science, I intend two things: science for its own sake and useful science." Weizmann also worked to raise the funds necessary to operate the Sieff Institute throughout its existence as an independent research institute.

For Weizmann's seventieth birthday, American Zionists founded a fundraising committee to raise money for a significant expansion of the Institute, which was to be renamed – with the consent of the Sieff family – the Weizmann Institute of Science. An impressive opening ceremony was held on November 2, 1949, just prior to Weizmann's seventy-fifth birthday. In addition to Weizmann himself, the speakers included Prime Minister David Ben-Gurion and Abba Eban. Yosef Sprinzak, speaker of the Knesset, was also in attendance. The Sieff Institute of Organic Chemistry became one of the new entity's five units, and it bears that name to this day. The leadership of the other units was offered to prominent scientists from around the world, but only one of them, Chaim Leib Pekeris, who became head of the Applied Mathematics department, accepted. Young scientists were appointed to head the three other units, initially on a temporary basis: Israel Dostrovsky, Isotopes; Aharon Katchalsky (Katzir), Polymers; and Ephraim Katchalsky (Katzir), Biophysics. The Experimental Biology department was established a bit later, when its chief, Isaac Berenblum, moved from England to Israel. In 1954 the Department of Nuclear Physics was established, headed by twenty-eight-year-old Amos de Shalit. The scientific faculty of the Weizmann Institute at its founding (beyond the faculty of the Sieff Institute) numbered forty – fourteen senior and twenty-six junior researchers. They quickly moved the Institute forward to become one of the world's most respected research institutions.

From its establishment, the Institute strove to operate in the spirit of Weizmann's scientific outlook. It purchased state-of-the-art equipment that enabled its scientists to do their work properly. Among the purchases was an analytic ultra-centrifuge. Purchased in the 1950s, it served the Polymer and Biophysics Departments. Another was Israel's first electron microscope, which, surrealistically, was brought to the Institute on a horse-drawn cart. Such items enabled the Institute's scientists to work in areas at the vanguard of science at the time. The Applied Mathematics department did the nearly impossible: In the space of a few years it constructed WEIZAC, one of the world's first electronic computers, with most impressive computing power and speed for its time. WEIZAC commenced operation in 1955 and lasted until 1963. Today it remains on exhibit in the building of the Faculty of Mathematics. Further advances in the field led to the construction of the Golem A and Golem B computers, which put the Institute at the forefront of computer science. Eventually, four Israelis, including three of the Institute's scientists, would go on to win the prestigious Turing Prize.

Today the Weizmann Institute is a multidisciplinary campus that is home to research in the natural sciences, from mathematics to biology. It is a research institute, not a university in the accepted sense of the word, as it does not offer undergraduate studies. However, it is an integral part of Israel's higher education system. About a thousand students are enrolled in its Feinberg Graduate School, studying for master's degrees and Ph.D.s. About 150 Ph.D.s are awarded each year, the largest number of such degrees granted in the natural sciences by any institution in Israel.

The Institute's central pursuit is basic research, in its five faculties: Mathematics, Physics, Chemistry, Biochemistry, and Biology. About half of the research is in the life sciences, where there are broad opportunities for cooperation among scientists of the different faculties in multidisciplinary and interdisciplinary research. The Institute also sponsors extensive educational activity. A relatively recent addition is the Davidson School of Science Education, which operates from the morning until late at night. It offers science instruction, training for science teachers, programs for outstanding students, and adult education.

But basic research, with a stress on excellence, remains the Institute's core activity. That can be seen in the success its scientists have had in winning prestigious grants. For example, Weizmann Institute scientists are members of thirty of the consortia participating in the European Union's FP7 program, and many of them have been awarded European Research Council (ERC) grants. Another measure of excellence is the high impact factor of the articles published by scientists at the Institute, placing it among the world's leading research facilities. Institute scientists collaborate with scientists around the world, from the United States to China, at an ever-growing rate. Nearly half of the scientific publications produced by the institute involve an overseas partner. The Weizmann Institute ranks high in worldwide ratings of research universities, right below the top ten. In this it has realized Chaim Weizmann's vision.

But Weizmann spoke explicitly not just about science and research for its own sake but also promoted applied work. The Weizmann Institute has played a pioneering role in technology transfer. To that end it established, in 1959, the Yeda Research and Development Company, long before similar companies were started at other universities in Israel and overseas. Yeda is responsible for identifying inventions produced by Weizmann Institute scientists that have commercial potential. It reserves intellectual property rights for the Institute and develops them to the transfer point, in cooperation with the relevant industries. Yeda exploits the Institute's scientific work in the development of pharmaceuticals, while earning royalties that support the institute's basic research program. Chaim Weizmann, it should be recalled, was a biotechnological innovator who in 1915, at the University of Manchester, isolated and developed a bacterium that produces acetone. In an act of technology transfer, he provided his discovery to the government, making an important contribution to the British war effort in World War I. Weizmann registered more than 120 patents during his lifetime.

Among the pharmaceutical products developed by Yeda have been Rebif, Copaxone, and Erbitux, and more are on the way. Nanotechnology products are already being marketed by large companies, and computer developments have also been commercially successful. Each year, tens of billions of dollars in sales ensue from products based on Weizmann Institute research.

The Weizmann Institute is indeed a living memorial to Chaim Weizmann, the founder of scientific research in Israel. It pursues basic research with an emphasis on excellence, while exploiting the practical applications of that research to the fullest. In doing so, it makes a vital contribution both to the progress of science and to the development of applications. On top of that, it has made an essential contribution to Israel's society and economy, just as its founder would have wished.

Benjamin Z. Kedar

The Description of Weizmann's First Encounter with Balfour, 1906

Legendary or Factual?

In present-day historiography one sometimes detects a tendency to cast doubt on the factuality of dramatic events that get star billing in history books. For example, a leading medievalist, Johannes Fried, has recently argued that the famous scene, alleged to have taken place in January 1077, of a barefoot, freezing, and contrite Emperor Henry IV being compelled to bang on the gates of Canossa Castle until Pope Gregory VII consented to forgive him, is a fiction; it appears solely in works written by supporters of the pope.[1] Similarly, though the deed figures in every book on the Reformation, there are historians who doubt that Martin Luther actually nailed his ninety-five theses to the church door of Wittenberg Castle in October 1517.[2]

How can this urge to de-dramatize historical occurrences be understood? We all have contrarian impulses at times, of course. But this phenomenon may also reflect some historians' displeasure with *histoire événementielle*, the history of events, and their turn toward a history of processes. Perhaps it also attests to the inability of some historians, spending their lives in unperturbed academic surroundings and rather

1 Johannes Fried, "Der Pakt von Canossa: Schritte zur Wirklichkeit durch Erinnerungsanalyse," in Wilfried Hartmann & Klaus Herbers (eds.), *Die Faszination der Papstgeschichte*, Cologne: Böhlau, 2008, pp. 133–197.

2 Erwin Iserloh, *Luthers Thesenanschlag: Tatsache oder Legende?*, Wiesbaden: Franz Steiner, 1962; Richard Friedenthal, *Luther: Sein Leben und seine Zeit*, Munich: R. Piper & Co., 1967, p. 155.

limited in their exposure to the outside world, to acknowledge the likelihood of a truly spectacular incident.[3] All these may be manifestations of deeper currents of which historians themselves are not necessarily aware, persuaded as they are that their conclusions are based solely on a detached appraisal of the facts.

In any case, one of the more stirring events in Chaim Weizmann's life, his encounter with Arthur James Balfour on January 9, 1906, has also undergone debunking. Weizmann, then thirty-one years old, had joined the faculty of the University of Manchester only a year and a half previously, and his English was still rudimentary. Balfour, aged fifty-nine, had stepped down as prime minister just a month earlier, on December 5, 1905. He visited the city while campaigning – unsuccessfully – to retain his seat for Manchester East in the House of Commons; he lost to the Liberal candidate four days later, part of a general rout of his Conservative party.

Weizmann relates in his autobiography that his meeting with Balfour was arranged by Charles Dreyfus, who had employed Weizmann as a research chemist in his capacity as director of a factory that produced aniline. Dreyfus was also chairman of the Manchester Zionist Society, chairman of the Manchester Conservative Party, and a supporter of the Uganda plan, which Balfour's government had proposed to the Zionist movement. He hoped that Balfour would persuade Weizmann that he had erred in resisting the plan to establish a Jewish homeland in East Africa. Balfour, for his part, wanted to speak to one of the Zionists who had opposed the plan.[4] When they met, Weizmann attempted to explain the opposition to the Uganda option, but the longer the conversation went on, the more apprehensive he became that Balfour was listening to him only out of politeness. "I felt that I was sweating blood and I tried to find some less ponderous way of expressing myself," he later wrote

3 The working routine of a historian who lives in a serene Western democracy is well described in a brilliant essay by Jack Hexter, "The Historian and His Day," in his *Reappraisals in History*, Evanston, IL: Northwestern University Press, 1961, pp. 1–13.

4 *Trial and Error*, p. 142.

in his autobiography. They exchanged several more sentences, when Weizmann suddenly said:

> "Mr. Balfour, supposing I were to offer you Paris instead of London, would you take it?" He sat up, looked at me, and answered: "But Dr. Weizmann, we have London." "That is true," I said. "But we had Jerusalem when London was a marsh."[5]

In his memoirs, Weizmann wrote that this conversation – his first with Balfour, by his account – marked the "far-off beginnings of a new chapter in the relationship between England and Zionism." The encounter was also a turning point in his own life, because the conversation with Balfour convinced him that he had to resume intensive Zionist activity. Yet he did not, he acknowledged, tell this story until many years later[6] – in a speech he made in Jerusalem in 1919.

But did Weizmann really pose to Balfour this pointed question about London and Paris, thus bringing home the importance to the Zionist movement of the ancient Jewish homeland? In the first volume of his authoritative biography of Weizmann, Jehuda Reinharz casts doubt on the accuracy of Weizmann's account of the meeting in his memoirs, which, he writes, "has about it all the ingredients of a good story," making it "only natural that it has been widely quoted in Zionist and general historiography." Reinharz himself presents the story in full but then immediately offers a caveat: "Weizmann's record of the meeting the very same day of the interview is a great deal less dramatic and, no doubt, more accurate." He quotes a letter written that day by Weizmann to his fiancée, Vera Khatzman:

> I had a meeting with Balfour today, and had a long and interesting conversation with him about Zionism. He explained that he sees no political difficulties in the attainment of Palestine – only economic difficulties. We spoke of territorialism [establishing a Jewish state in

5 *Ibid.*, p. 144.
6 *Ibid.*, pp. 142, 145.

> a territory other than Palestine]. I explained to him why this was not possible. We undertook to send him a memorandum.[7]

In fact, the report to Vera on the day of the meeting does not contradict the account Weizmann included in his memoirs many years later. In all probability, the discussion of territorialism focused on the Uganda proposal, and the long and interesting conversation on Zionism could certainly have included Weizmann's riff on London, Paris, and Jerusalem. Reinharz adduces sources that shore up his doubt about the factual basis for the "good story," but his certainty that the less dramatic description is more precise perhaps hints at a skeptical attitude toward the factuality of dramatic accounts. He also questions Weizmann's claim that the meeting was prompted by Balfour's interest in speaking with a Zionist who had played a role in bringing about the rejection of the Uganda plan. He conjectures that Balfour, at the climax of his election campaign, hoped that Weizmann might influence Jewish voters in his favor. Balfour was worried about the Jewish vote, because his government had, just a year before, brought about the passage of the Aliens Act, which was justly perceived as aimed at restricting Jewish immigration to England. According to Reinharz's conjecture, Balfour stressed the Uganda plan in his conversation with Weizmann in the hope of persuading Jewish voters that, despite the Aliens Act, his government was not hostile to Jewish interests.[8]

Let's take a look at the evidence that Reinharz claims justifies his skepticism about Weizmann's later version of his conversation with Balfour.

The principal exhibit in Reinharz's claim regarding the dubious factuality of the "London and Jerusalem" anecdote is that Weizmann makes no mention of it in his letter to his fiancée, written that same day. It would not have been in Weizmann's character, Reinharz argues, to

7 Jehuda Reinharz, *Chaim Weizmann: The Making of a Zionist Leader*, New York–Oxford: Oxford University Press, 1985 (= Reinharz, *CW*, I), pp. 271–272; Weizmann to Vera Khatzman, *CWL*, IV, no. 195, p. 219.

8 Reinharz, *CW*, I, pp. 272–275.

downplay the significance of such an important meeting. On the contrary, elsewhere he made much of meetings with far less important figures, such as Lord Percy and Sir Clement Hill of the Foreign Office. In other words, if his conversation with Balfour, which Weizmann "purported to remember many years later," was as interesting as he presented it in his autobiography, it is only reasonable to presume that he would have reported it faithfully to Vera. Yet he did not mention it in public before his speech in Jerusalem thirteen years later.[9]

But, as is well known, an *argumentum ex silentio*, based on the absence of evidence, is problematic. Only when there is convincing reason to believe that a certain event should indubitably have been mentioned can the fact that it was not mentioned be taken to suggest that the event did not occur; and even then it is no more than a hypothesis. In the case at hand, can we really be so certain that Weizmann, in his letters to his fiancée, reported all the events and conversations he was having at the time? It seems sufficient to note that on January 14, 1906, Weizmann took part in the annual convention of the English Zionist Federation in Glasgow, where he was elected to the executive board for the first time. But his letter to Vera that day makes no mention of this significant new office.[10] Consider also that Balfour, by the time Weizmann met him, was no longer as important as he had been. True, he was a former prime

9 *Ibid.*, p. 272. Although Reinharz does not mention it, an earlier biographer, Barnet Litvinoff, had noted nine years previously that Weizmann did not recount the exchange on London and Jerusalem in his letter to Vera, even though he reported almost verbatim his conversations with lesser personages such as Percy and Hill. Litvinoff also linked Balfour's interest in meeting Weizmann to Balfour's problems with Jewish voters as a result of the Aliens Act. While leaving open the question of what exactly what was said at the meeting, he reasoned that the stories reported by Blanche Dugdale in her biography of Balfour, and later by Weizmann in his autobiography, were recast and polished for propaganda purposes. See Barnet Litvinoff, *Weizmann: Last of the Patriarchs*, London: Hodder & Stoughton, 1976, p. 61. Reinharz includes Litvinoff's book in his bibliography. See also Leonard Stein, *The Balfour Declaration*, Jerusalem–London: Magnes Press, 1961, pp. 148–151.

10 *CWL*, IV, no. 197, p. 221. In fact, Reinharz notes this anomaly: "It is curious that his election did not even seem important enough to report to Vera." Reinharz, *CW*, I, p. 278.

minister, but it was clear to all that he stood on the brink of defeat. Indeed, before mentioning the meeting in his letter to Vera, Weizmann informed her: "There is great excitement here over the elections. Balfour's chances are slender."[11] Balfour did lose the election, and the years that followed were not good ones for him. In 1911 he resigned as leader of the opposition. It is thus not surprising that Weizmann made no mention of that conversation in public for over a decade.

Balfour, in his turn, did not forget the conversation. When in the fall of 1914 Weizmann asked his friend Samuel Alexander, a professor of philosophy at Manchester University who knew Balfour well, to arrange a meeting with him, Balfour replied to Alexander on November 17: "I have the liveliest and also the most pleasant recollections of my conversation with Dr. Weizmann in 1906, and should be very happy to hear from him."[12] After the two men met in London on December 12, Weizmann reported to Ahad Ha'am: "Balfour remembered everything we discussed eight years ago, and this made it unnecessary for me to explain once again the nationalist formulation of the Jewish question."[13] About a month later, in a report to the members of the Executive of the International Zionist Organization, Weizmann wrote in the same vein: "I may be allowed to mention at this point that I had the privilege of talking to Mr. Balfour about Zionism 8 years ago, and that I had reason to believe that Mr. Balfour remembered this conversation."[14]

Weizmann cited the 1906 conversation in public in the Hebrew speech he delivered in Jerusalem on December 17, 1919. By that time, Balfour's fortunes had turned: he had served as foreign minister during the latter half of World War I and had, in that capacity, issued the Balfour Declaration.

The Jerusalem speech is worth examining. (Reinharz refers to the text

11 *CWL*, IV, no. 195, p. 219.

12 Quoted in Stein, *The Balfour Declaration* (above, note 9), p. 153.

13 Weizmann to Ahad Ha'am, *CWL*, VII, no. 68 (December 14, 1914), p. 81.

14 Weizmann's report of January 7, 1915, *CWL*, VII, no. 95, p. 114. See also Jonathan Schneer, *The Balfour Declaration: The Origins of the Arab-Israeli Conflict* (New York: Random House, 2010), p. 134. My thanks to Prof. Moshe Zimmermann for having brought these texts to my attention.

included in the first volume of Weizmann's writings, published in 1936 in Tel Aviv,[15] while I, aware of the tendency of Zionist leaders, like many others, to spruce up their words before putting them into books, have preferred the original, "unrefined" version published in the Hebrew newspaper *Ha'aretz* on December 19, 1919, two days after the speech was given.) By and large, Weizmann recounted the conversation in his speech much as he did thirty years later in his autobiography. Balfour wanted to hear why he opposed the Uganda plan; Weizmann had difficulty persuading him, until he had the idea of posing his query about London and Paris. But in the speech, his presentation is much more outspoken. Weizmann presented the conversation as "the bridge toward the recent political negotiations," not as the "far-off beginnings of a new chapter in the relationship between England and Zionism," as he wrote in his autobiography.

According to the speech, Weizmann asked Balfour: "Would you trade London for the more beautiful Paris?" In the autobiography, as we have seen, he phrased himself more tactfully: "Supposing I were to offer you Paris instead of London, would you take it?" In his speech, Weizmann recalled that Balfour's response was forceful: "No! But London is ours and Jerusalem is not yours!" The autobiography has Balfour saying "But Dr. Weizmann, we have London." In the autobiography, Weizmann declared: "But we had Jerusalem when London was a marsh," whereas in the speech, he recalls that he stated bluntly: "Jerusalem is ours. We developed a world culture there when your ancestors still walked as savages on this island called England."[16] It should also be noted that in his speech Weizmann erred in referring to Balfour as, at the time of their meeting, "the prime minister who presented the Uganda proposal to the Zionists." He perhaps did not recall

15 Reinharz, *CW*, I, p. 487, note 84; Chaim Weizmann, *Words: Speeches, Articles and Letters*, I, Tel Aviv: Mitzpe, 1936, pp. 133–137 (Hebrew). The relevant passage appears on p. 134.

16 *Ha'aretz*, December 19, 1919, p. 2. In the 1936 book (above, note 15), the final sentence is restyled in Hebrew: "We developed a world culture there at a time when your ancestors still lived like savages on this island called England."

that by the time they met Balfour was already out of office, or perhaps he thought it better to make Balfour out to have been more important at the time than he actually was.[17]

Weizmann's statement regarding a "world culture" that developed in Jerusalem "when your ancestors still walked as savages on this island called England" may well have been influenced by a retort attributed to Benjamin Disraeli: "Yes, I am a Jew, and when the ancestors of the right honorable gentleman were brutal savages in an unknown island, mine were priests in the temple of Solomon." Disraeli is presumed to have said this in 1835 to the Irish leader Daniel O'Connell, when the latter disparaged his Jewish origins,[18] but there is no written source to confirm it.[19] True or not, a slightly different version of the quote was cited by an American clergyman in 1903, and an important Jewish publisher presented it as authentic.[20]

In his speech in Jerusalem, Weizmann stated: "Balfour mentions this conversation in his introduction to Sokolow's book." While Reinharz notes that Balfour did mention the exchange in his introduction to Nahum Sokolow's *History of Zionism*, published in 1919, he claims that Balfour did not quote what was actually said.[21] This is only partly true. In the introduction, dated September 20, 1918, Balfour relates how he had supported the Uganda plan. While his intentions had been good,

17 In his autobiography, Weizmann also errs in stating that, at the time of their meeting, Balfour was competing for the parliamentary seat of the "Clayton division of North Manchester"; see *Trial and Error*, p. 142. In fact, he stood for Manchester East; Winston Churchill stood for (and won) the seat for northwest Manchester.

18 See, for example, John Bartlett, *Familiar Quotations*[15], Boston: Little, Brown and Co., 1980, p. 501, no. 9; Geoffrey Wigoder, *Dictionary of Jewish Biography*, New York: Simon and Schuster, 1991, p. 119; Ivan Kalmar, "Jewish Orientalism," in Judith Targarona Borrás & Angel Sáenz-Badillos *(eds.), Jewish Studies at the Turn of the Twentieth Century: Proceedings of the 6th EAJS Congress, Toledo, July 1998*, Leiden: Brill, 1999, p. 308.

19 Burton Stevenson, *Stevenson's Book of Quotations, Classical and Modern*[8], London: Cassell, 1956, p. 1011, no. 14.

20 Cyrus Adler (ed.), *The Voice of America on Kishineff*, Philadelphia: Jewish Publication Society of America, 1904, p. 229.

21 Reinharz, *CW*, I, p. 487, p. 84.

the plan suffered from a fundamental flaw. "It was," he acknowledged, "not Zionism." It tried to find the Jews a home far distant from the land in which their faith had come into the world. "Conversations I held with Dr. Weizmann in January, 1906, convinced me that history could not thus be ignored, and that if a home was to be found for the Jewish people, homeless now for nearly nineteen hundred years, it was vain to seek it anywhere but in Palestine."[22] Thus, Balfour did refer to the content of the 1906 conversation: The context was the Uganda proposal, and Weizmann succeeded in persuading him that there was no Zionism without the Land of Israel.

Now, according to Reinharz, when Weizmann composed his memoirs at the end of the 1940s, another already extant book offered the main points of the story of his conversation with Balfour, as Weizmann himself told it.[23] This is true, but it requires a closer look. The book in question, published in 1936, is the biography of Balfour written by his niece Blanche Dugdale, an enthusiastic Zionist and a member of Weizmann's inner circle.[24] A comparison of this volume's account of the conversation – which Dugdale presents as a direct quote from Weizmann – with Weizmann's version of thirteen years later shows that they are indeed much alike, and sometimes literally identical:

> I began to sweat blood to make my meaning clear through my English. At the very end I made an effort, I had an idea. I said: "Mr. Balfour, if you were to be offered Paris instead of London, would you take it? Would you take Paris instead of London?" He looked surprised. He: "But London is our own!" I said: "Jerusalem was our own when London was a marsh."[25]

22 Arthur J. Balfour, "Introduction," in Nahum Sokolow, *History of Zionism, 1600–1918*, New York: Ktav, 1969 (1919), pp. xlix–l.

23 Reinharz, *CW*, I, p. 272.

24 Norman A. Rose (ed.), *Baffy: The Diaries of Blanche Dugdale, 1936–1947*, London: Vallentine Mitchell, 1973.

25 Blanche E.C. Dugdale, *Arthur James Balfour, First Earl of Balfour*, I, London: Hutchinson, 1936, p. 434.

Weizmann may well have given Dugdale his account of the conversation in writing, and then polished it up a bit for use in his autobiography. Another possibility is that the story had taken form in Weizmann's memory by the 1930s, and that he recounted it from time to time.

Dugdale wrote that her uncle had spoken to her frequently of the impression made on him by his conversation with Weizmann, and by the latter's firm rejection of the Uganda offer.[26] Furthermore, in an essay on Weizmann and Balfour that she published in 1944, she related that she had heard about the conversation from Balfour a short time after it took place, and from Weizmann years later. As she recalled, when her uncle returned from Manchester following the elections of January 1906 to spend a brief vacation with his family – she was twenty-six at the time – he also held forth about Zionism and the impression its spokesman, Weizmann, had made on him. He brought this up even though there were much more pressing matters on his mind at the time, most centrally his election defeat. The conversation was so fateful, in her view, that when she sat down to write her uncle's biography thirty years later, she went so far as to visit Manchester's Queens Hotel and locate the room in which it had taken place, making a veritable pilgrimage to a Zionist *lieu de mémoire*.[27] Moreover, Balfour had once rehearsed for her a part of the exchange with Weizmann as he recalled it: "'Would you change London for Paris, Mr. Balfour?' 'No – but London is the capital of my country.' 'Jerusalem was ours, when London was a marsh.'"[28] Reinharz,

26 *Ibid.*, p. 435. Her quote from Balfour was recorded after a conversation with him on August 7, 1929. See Stein, *Balfour Declaration* (above, note 9), p. 152, note 29.

27 Blanche Dugdale, "Weizmann and Balfour," in Meyer W. Weisgal (ed.), *Chaim Weizmann: Statesman, Scientist, Builder of the Jewish Commonwealth*, New York: Dial Press, 1944, pp. 131–132. While Balfour was most impressed by Weizmann, he did not remember his name. Some time after the encounter, he asked Dreyfus to provide him with the name of the "Russo-Jewish gentleman" with whom he had spoken in Manchester. See Stein, *Balfour Declaration* (above, note 9), p. 152.

28 Dugdale, "Weizmann and Balfour" (above, note 27), p. 132. In 1963, Israel Sieff, a member of Weizmann's inner circle since 1913, repeated this version almost verbatim; see Israel M. Sieff, "The Manchester Period," in Weisgal, *Chaim Weizmann* (above, note 27), p. 93.

who is acquainted with Dugdale's 1944 essay,[29] does not mention that it quotes part of the 1906 exchange according to Balfour's recollection, as recorded by his niece, in a version quite similar to that offered by Weizmann in his autobiography.

This brings us to the last piece of evidence regarding the Weizmann–Balfour conversation, the testimony of William Emmanuel Rappard. Reinharz addresses it in a long footnote, which he begins by acknowledging that the version offered by Weizmann in his memoir is supported by Rappard, who, as a member of the League of Nations Permanent Mandates Commission, in 1925, witnessed an encounter between Balfour and Weizmann at which they recalled their earlier meeting in Manchester. Later in the footnote, Reinharz raises serious reservations about Rappard's report: "It should be kept in mind, however, that this conversation took place twenty years after the event, and that Rappard recorded it for the first time almost thirty years after hearing it, i.e., after *Trial and Error* had already been published. What makes his account suspect is the fact that it matches Weizmann's own wording and style."[30]

Reinharz's argument is not cogent, however. He bases his footnote on an article by Rappard that was published in English on October 30, 1953, in a weekend supplement to the *Jerusalem Post* devoted to Weizmann's memory. Yet Rappard states unequivocally in this article that when he sailed in 1925 to take part in the Hebrew University's opening ceremony, he had the privilege of sitting between Weizmann and Balfour while they recalled their 1906 meeting – and that he, Rappard, recorded their words "on the spot,"[31] not almost thirty years later, as Reinharz claims. Furthermore, Reinharz seems to be unaware of the full version of Rappard's account, which he published in January 1953 in Switzerland under the title *A la mémoire de Chaim Weizmann*. There,

29 Reinharz, *CW*, I, p. 487, note 89.

30 *Ibid.*, note 82.

31 William E. Rappard, "Great Humanitarian," *The Jerusalem Post: Chaim Weizmann Memorial Supplement*, October 30, 1953, p. 1.

too, Rappard emphasizes that he took down the exchange between Weizmann and Balfour right upon (*immédiatement*) hearing it.[32]

Who was William Emmanuel Rappard, and can he be taken as a reliable witness? Born in New York in 1883 to Swiss parents, he studied at Harvard and at the University of Vienna, taught political economy at Harvard, and from 1913 onward was professor of economic history at the University of Geneva. He wrote widely on history and current events in major European and American learned journals. During World War I he was active in the International Red Cross and carried out diplomatic missions, among others to President Woodrow Wilson. In 1920 he was appointed secretary of the Permanent Mandates Commission of the League of Nations, and he served as a member of that body from 1924 to 1939.[33] In that capacity he was involved with the British Mandate in Palestine and from 1921 onward had a working relationship with Weizmann, which evolved into a friendship. Weizmann expresses great respect for Rappard in his autobiography.[34]

It is hardly credible that a historian like Rappard would have fabricated his account of the on-board conversation between Weizmann and Balfour, and there is no reason to doubt his claim that he wrote it down at the time. I have not found his actual transcription of the conversation among his papers, which are deposited in the Federal Archives of Switzerland in Bern, and it does not appear in the diary he kept on his trip to Palestine.[35] Presumably he wrote it down elsewhere. However, the

32 William E. Rappard, *A la mémoire de Chaim Weizmann, principal fondateur et premier président de l'État d'Israël, ami de la Suisse*, Neuchâtel: Éditions de la Baconnière, 1953, p. 35.

33 See Victor Monnier's comprehensive biography, *William E. Rappard: Défenseur des libertés, serviteur de son pays et de la communauté internationale*, Geneva–Basel: Edition Slatkine & Helbing et Lichtenberg, 1995. Rappard's 1925 trip to Palestine is recounted on pp. 335–346.

34 *Trial and Error*, pp. 463, 465.

35 The only reference Rappard makes in his journal, recorded on board the SS *Esperia* on March 21, 1925, is "Conversation extraordinairement vive entre B., W. + moi," See Bern, Schweizerisches Bundesarchiv, Nachlass William Rappard, J1.149 1972 135, Band Nr. 120-202 – Mission à Jérusalem, 1925.

archives contain the drafts of Rappard's 1953 book. A comparison of the drafts indicates that in several places he raised the linguistic register of the original account of the conversation.[36]

Had Rappard read Weizmann's autobiography by the time he wrote his book? Certainly.[37] A lengthy section consists of passages from *Trial and Error*, principally regarding Switzerland.[38] But is there any basis for the claim that the 1925 reconstruction by Weizmann and Balfour of their 1906 conversation was affected by the wording and style of Weizmann's version in his autobiography? Definitely not. First, it is chronologically impossible. Furthermore, in his autobiography, as already noted, Weizmann writes that the meeting with Balfour was initiated by Dreyfus, a Conservative and a Zionist, and that Balfour was interested in meeting a Zionist who opposed the Uganda plan. Rappard's account of the 1925 conversation[39] begins, however, with Balfour asking Weizmann how he had managed to gain an audience with him, and Weizmann clarifies that he himself had asked to meet Balfour because he had wanted to speak to him about Zionism, and that he had managed to get his name onto a list of people whom Balfour was to meet during his visit.

Weizmann goes on to say that he had carefully prepared his brief presentation, but when he made it he saw that he had not managed to gain Balfour's attention. Rappard has Balfour commenting: "Indeed, I did not understand the point of the lecture and my head was full of more urgent matters." Rappard's account thus amounts not just to a recreation of the exchange, but also to an attempt by the two men to recall what each was thinking as it went on. That element is certainly not present

36 A comparison between the article published in the *Jerusalem Post* and the relevant passage in the French book also reveals that the article renders the French original inaccurately.

37 Incidentally, a copy of *Trial and Error* bearing Weizmann's dedication to Rappard, dated June 2, 1949, was offered for sale on site www.universityarchives.com (accessed January 2, 2014). When this article went to press, the item was no longer listed on the site.

38 Rappard, *A la mémoire* (above, note 32), pp. 16–27.

39 See the Appendix.

in Weizmann's autobiography or in the other accounts. Furthermore, the incisive question posed by Weizmann in order to get Balfour to take notice appears in a version unlike that of all the other accounts: "Mr. Balfour, what do you prefer, Paris or London?" To which Balfour replies: "Yes, I recall that question, which seemed quite absurd." The bottom line is that there is no basis for the claim that Rappard was influenced by the account of the 1906 meeting in Weizmann's autobiography. Rather, he presumably decided to make public in 1953 the record he had put down in 1925, in order to offer an alternative to the version in Weizmann's book. In doing so, he also provided a rare testimony to the respect Weizmann paid to Balfour in an informal context.

So who initiated the 1906 meeting – Dreyfus and Balfour, as the autobiography states, or Weizmann, as Rappard's account indicates? My inclination is to ascribe greater accuracy to Rappard's version. It is corroborated by the account of Harry Sacher, a member of Weizmann's circle of Zionists in Manchester, who wrote in 1944 that Weizmann contacted Balfour "on his own initiative," and that this led to their meeting. Sacher made a similar statement after Weizmann's death as well, and there is no reason to doubt its veracity.[40]

If that is the case, why did Weizmann write his autobiographical account as he did? There are several possible explanations, but the following one seems simplest. The meeting in January 1906, which Weizmann and others presented as the first between him and Balfour, was in fact their second. In a letter he wrote to Vera a year earlier, on January 28, 1905, Weizmann mentioned that he had the previous day attended a public meeting at which Balfour spoke, and that he had managed to speak to the prime minister about Zionism for about five minutes. Balfour, Weizmann told his fiancée, had promised to set aside a longer time for a meeting in the future.[41] The editors of the volume in which

40 Harry Sacher, "The Manchester Period," in Weisgal, *Chaim Weizmann* (above, note 27), p. 188; Harry Sacher, "Chaim Weizmann," in idem, *Zionist Portraits and Other Essays*, London: Anthony Bloud, 1959, p. 7.

41 *CWL*, IV, no. 15, p. 19. Cf. *ibid.*, no. 14, p. 18; no. 16, p. 21. And see Reinharz, *CW*, I, p. 244.

this letter was published believe that this first meeting was set up by Dreyfus,[42] and that sounds reasonable. Weizmann, writing his memoirs many years later, seems to have merged the two encounters into one.

In short, I see no reason to doubt the factual nature of Weizmann's later account of his 1906 exchange with Balfour. Whatever the precise words they used, London, Paris, and Jerusalem all appeared. The conversation deserves to be restored to its rightful place in the history of the Zionist movement.

Appendix:
The account of the Weizmann–Balfour conversation in the draft of William Rappard's *A la mémoire de Chaim Weizmann* (1953)

The draft is preserved in the Federal Archives of Switzerland in Bern, where it is referenced as Nachlass William Rappard, J1.149, Band Nr. 103 – Manuscrits Rappard: e.a. Chaim Weizmann, Varia politica 1953–54. The file contains the original typescript of the book, marked as version III, with corrections in ink. The published volume reflects these corrections, as well as a few other minor corrections that are not attested in the file.

I offer here the earlier typescript version (pp. 20–21 of version III); the changes in the text that appear in the book itself (pp. 35–37) are indicated in my notes. The text before and after this section, which Rappard presumably based on his diary, reflect what he knew at the time he wrote the book.

Le voyage de Gênes à Alexandrie fut du reste mémorable à tous égards. A Naples nous fûmes rejoint par M. Balfour qui, accompagné de son neveu et de sa nièce Lascelles, se rendait à Jérusalem pour y inaugurer l'Université. Balfour, qui n'avait pas le pied marin, s'était pour cette raison dispensé de s'embarquer à Gênes. Heureusement qu'en quittant

42 *CWL*, IV, p. 25. Stein said as much previously, in idem, *Balfour Declaration* (above, note 9), p. 147.

Naples[43] la mer, qui devait se troubler bientôt, était encore calme. Je dois à cette circonstance fortuite d'avoir passé une des soirées les plus intéressantes de ma vie. Après le dîner, en effet,[44] nous fumes réunis, les Balfour, les Weizmann et moi, dans le salon. Il s'y engagea, entre Balfour et Weizmann, une conversation dont il n'est nullement grandiloquent de la déclarer historique. En effet, tout le monde sait que c'est en 1906 que Weizmann convertit Balfour au sionisme.[45] Weizmann était alors un jeune et obscure chimiste et Balfour le Premier Ministre de Sa Majesté britannique[46] qui venait dans sa circonscription électorale pour s'y faire réélire. Or la conversation qui s'était, il y avait près de vingt ans auparavant, engagée entre Weizmann et Balfour fut reconstituée par les deux interlocuteurs dans des termes que j'ai notés immédiatement après les avoir recueillis:

> Comment, demanda Balfour, avez-vous obtenu le rendez-vous qui m'a valu le plaisir de faire votre connaissance?
> **Weizmann:** Je me disais que l'occasion de votre venue à Manchester était sans doute providentielle. Plus qu'aucun autre homme d'Etat je tenais, en effet,[47] à vous intéresser à notre cause en raison même de vos fonctions et aussi de ce que je croyais être votre attitude probable envers elle. J'obtins donc, à force de démarches, je l'avoue, d'être inscrit dans la liste à vrai dire très longue de vos rendez-vous. Vous avez bien voulu me recevoir et écouter mon plaidoyer.[48] J'avais soigneusement préparé mon petit discours, mais malgré la courtoisie avec laquelle vous fîtes semblant de m'écouter, je me rendis bien compte que je ne réussissais pas à capter votre attention.
> **Balfour:** En effet, je ne comprenais pas l'à propos de cet exposé et j'avais la tête pleine de préoccupations plus urgentes.

43 qu'au large de Naples
44 "En effet" is omitted in the book.
45 This is, of course, the popular wisdom about the conversation, but it has no basis in reality.
46 As noted above, Balfour was no longer prime minister at the time.
47 "En effet" is omitted in the book.
48 Instead of *écouter mon plaidoyer*, the book has *prêter l'oreille à mon plaidoyer*.

> **Weizmann:** Comme je voyais s'écouler le nombre très limité de minutes qui m'avait été accordé, je me résolus à jouer le tout par le tout. Désireux de capter[49] à tout prix votre attention, je vous demandai donc à brûle-pourpoint: Monsieur Balfour, qu'est-ce que vous préférez, Paris ou Londres?
> **Balfour:** Oui, je me souviens de cette question qui m'avait parue très saugrenue.
> **Weizmann:** Je la hasardai pour me fournir l'occasion de vous dire, à la suite de la réponse que j'escomptais, ce qui suit.[50] Comme j'avais osé le prévoir, vous me dites en effet:[51] Mais c'est[52] évident que je préfère Londres, parce[53] que Londres est à nous. C'est alors que j'osais vous dire:[54] Monsieur Balfour, est-ce que vous vous rendez compte que Londres était un marais quand Jérusalem était à nous?
> **Balfour:** Oui, je me souviens de cette déclaration aussi impressionante qu'elle était inattendue.

Ainsi que je le savais par ailleurs, ç'avait été le point de départ d'une longue conversation qui, malgré l'impatience manifestée par les secrétaires du Premier Ministre,[55] se prolongea pendant près d'une heure. Lorsqu'elle fut terminée, Balfour été gagné au sionisme.[56]

Cette conversation de l'Esperia[57] me restera présente à l'esprit tant que je vivrai. C'est un privilège extrêmement rare, en effet, pour un historien de pouvoir vérifier ainsi l'exactitude d'un récit qu'il avait

49 Rappard struck out the word *capter* and inserted *forcer*, which was printed in the book.
50 Rappart struck out *ce qui suit* and inserted *ce qui devait vous frapper*, which was printed in the book.
51 "En effet" is omitted in the book.
52 C'est
53 Rappard struck out *parce* and inserted *mais c'est*, which was printed in the book.
54 Instead of *que j'osai vous dire*, the book has *je me risquai à vous demander*.
55 See above, note 46.
56 See above, note 45.
57 The boat on which the group sailed to Alexandria.

souvent eu l'occasion de lire auparavant.[58] Balfour et Weizmann, sans jamais se contredire, complétaient leur récit par des remarques et des commentaires bien faits pour donner un relief accru à cette conversation assurément étonnante.

58 Rappard clearly could not have read about the conversation prior to his boat trip in 1925. Perhaps he was referring to his feeling when he compared his notes from 1925 with the descriptions of the conversation that appeared later.

Shlomo Avineri

Weizmann: The Making of a Statesman

Theodor Herzl left Constantinople for the last time in June 1902. All his attempts to obtain a charter from the sultan for a Jewish homeland in Palestine had failed. On the long train journey to Vienna, he set down in his diary a detailed account of his efforts, stage by stage, seeking to understand why he had failed and whether the entire idea of negotiating with the Ottoman government had been a mistake.[1] Noting the deceptions, intrigues, and corruption that pervaded the Sublime Porte, he tried to console himself. At least, he wrote, "I have left the cave of Ali Baba and the forty thieves." Fundamentally, though, he remained convinced that, in the long run, the key to achieving Zionism's goals would be found in the Turkish capital, even if his attempts so far had been premature. Perhaps they would succeed once the Ottoman regime lost a future war. Bulgaria, he noted, had obtained independence from the Turks following the Russo-Turkish war of 1877–1878.

The Balfour Declaration of 1917 was in a sense a vindication of this prediction, even if Herzl could not have foreseen the historical context in which it occurred. Nevertheless, while the efforts to obtain the declaration and the steps that followed it carried on Herzl's view of political Zionism, the achievement was not simply a linear extension of that concept. Rather, it required an essential reorientation of Zionist policy. That was accomplished, to no small extent, without any explicit mandate from the Zionist movement's institutions, which found themselves sidelined in the wake of the disruptions caused by World War I. The process

1 See my book, *Herzl: Theodor Herzl and the Foundation of the Jewish State*, London: Weidenfeld and Nicolson, 2013, pp. 201–202.

came to fruition only thanks to the novel and fluid circumstances that the war brought on, and to a confluence of events that no person controlled.[2] In any case, Zionism's first significant political success was clearly bound up with far-reaching world historical events that, while not deriving from developments connected to the Jewish people itself, affected Jewish history no less than the efforts and desires of Jewish figures and organizations. Here, too, an ability to exploit historical opportunity, and to ride the waves of what the Greeks called καιρός (*Kairos*), and what Machiavelli, in a different philosophical sense, referred to as *fortuna*, proved themselves in the face of quasi-deterministic historical explanations that cast every event as preordained. Such views attach mounds of explanations to the rational intentions of various individuals while disregarding the chance, arbitrariness, and lack of internal logic that are so much part of historical events.

The Zionist Organization did not collapse in the wake of Herzl's death in 1904, but his loss contributed to its ongoing marginality. Herzl's personality had fired the imaginations of masses of Jews and imbued them with almost messianic fervor. While his political and diplomatic efforts did succeed in bringing the Zionist idea to the attention of political leaders and into international discourse, they had in the end led nowhere. This could not help but evoke in some minds the specter that opponents of Zionism had already brought up during his lifetime – that of Shabbetai Zvi, the false messiah of the seventeenth century. That Herzl had left no successor who matched his stature left the movement dispirited. David Wolffsohn was not Herzl, and the Uganda crisis also left a bad taste in its wake, not to mention much disappointment. Zionist Congresses continued to convene on a nearly annual basis; and the movement made a series of significant land purchases in Palestine, thanks largely to the quiet and persistent work of Arthur Ruppin, director of the Zionist Organization's Palestine Office in Jaffa. A significant

2 That confluence has never been highlighted better than on the front page of the *Times* of London on November 9, 1917, on which the headlines of two side-by-side articles read: "Coup d'Etat in Petrograd, Lenin Deposes Kerensky – Peace and Bread"; "Palestine for the Jews – Official Sympathy."

number of Zionist pioneers did indeed arrive in Palestine in the Second Aliya, and the foundation of Tel Aviv, the "first Hebrew city," warmed the hearts of many. But diplomatic activity aimed at granting political, if not sovereign, status to the idea of the Jewish return to Zion dissipated as the energy Herzl had harnessed dispersed in other less significant directions. The founder's achievements may have been minimal, but his dynamism had moved the Zionist idea from the margins to the center, giving it standing in both intra-Jewish discourse and the international arena. The decade between his death and the outbreak of World War I was, however, characterized by a low level of activity devoid of any political efforts or achievements. While the Zionist Yishuv, that section of the Jewish community in Palestine committed to the Zionist idea, grew slowly despite the obstacles the Ottomans placed in the way of Jewish immigration, that could hardly make up for the movement's political ineffectiveness.

World War I shook the small movement in an unprecedented way, just as it did Europe and the Jewish people as a whole. A radically different international constellation emerged from the chaos. Weizmann's contribution to channeling these new circumstances toward the goals of Zionism was significant, but it needs to be examined critically. That is especially the case in light of the historiographical tendency, which Weizmann proved very good at promoting and even inflating, to amplify his actual achievements, even though they are impressive in and of themselves.[3]

The outbreak of war paralyzed the Zionist movement. Germany and Russia, the two countries that, each in a different way, were the movement's main theaters of activity, were enemies. The Zionist Organization, which had its administrative center in Germany (Otto Warburg

3 On Weizmann's personal attributes, see especially Isaiah Berlin, *Personal Impressions*, ed. Henry Hardy, Princeton: Princeton University Press, 2001, pp. 44–65. In many ways, Weizmann became Herzl's first worthy successor, as perceived both from outside the movement and from within the Jewish public. He also succeeded in deepening the connection between the movement and east European Jewry, a group that, for Herzl, was largely alien. Weizmann was uniquely able to present himself as both an English gentlemen and a Jew from the *shtetl*.

was its president when the war broke out), found itself cut off from most of its members and activists in Russia and in the territory of Congress Poland, part of the Russian empire. In other words, the leadership lost contact with its largest base of supporters. Under the circumstances, the movement's decision to remain neutral is understandable. It sought to maintain its ties with its members on both sides of the battle lines via its office in Copenhagen. Tactically, this was the right decision, but strategically it was a blunder. Given that the fate of the world stood in the balance and would be determined by the outcome of the war, Zionist neutrality meant Zionist paralysis.

On top of that, the Yishuv was dealt a heavy blow when the Ottoman government moved to deport all Russian nationals – including most of the Jews who had arrived in the Second Aliya – from Palestine, emptying it of most of the Zionist members of the Jewish population (the members of the religious, non-Zionist Old Yishuv in Jerusalem were for the most part Austro-Hungarian and German subjects). Ruppin in Jaffa, like Richard Lichtheim, the Zionist Organization's representative in Constantinople, were able, in their quiet and effective ways, to stave off economic collapse and starvation among Palestine's Jews. To this end they harnessed philanthropic contributions, mostly from the United States, conveyed to Palestine via Copenhagen. Figures like David Ben-Gurion and Vladimir Jabotinsky, then at the beginning of their political careers, understood that Zionism could not sit out the war and had to choose sides. Each of them pushed, in his way, to further ties with Britain, but with little effect.

This was the vacuum Weizmann stepped into. Up to this point, he had been a respected but not a central member of the Zionist movement. When the war broke out he was one of two vice-presidents of the English Zionist Federation but was primarily focused on furthering his scientific research career, not an easy task for a Russian-Jewish immigrant in England's rather cliquish academic community. It is no coincidence that he was employed largely in applied and industrial research. The paradox was that, from his location in provincial Manchester, he gradually fashioned a new political orientation that helped him extricate the Zionist Organization from the doldrums and launch it on a new political path. At the start, he did this without any sort of mandate from the Zionist Organization itself, and in the face of internal power struggles (for example, with Nahum Sokolow, from 1911 a member of the Zionist Executive, who was one of the few Russian Zionists who found himself in the West during the war). Nevertheless, when the Balfour Declaration was issued, Weizmann was not seen as its sole progenitor, partly because most of the talks leading up to it had been secret. He rose to head the movement only later, during the struggle for the political future of Palestine. After all, it was only at the 1920 San Remo conference that the powers finally agreed that Palestine would come under British rule, and only in 1922 did the League of Nations grant Britain a mandate to rule Palestine that incorporated the British commitment to the Zionist movement contained in the Balfour Declaration – without which the Declaration would have been worthless. In 1917, when the Declaration was issued, it was not at all clear that Britain would gain control of the country and stay on there as its ruler; things could have developed differently. Weizmann's diplomatic struggle to ensure that this happened is much less known than his role in achieving the declaration, but it is at least as important.[4]

The Balfour Declaration was not only the Zionist movement's first significant political and diplomatic achievement. It also redeemed Herzl's name and memory from the charge of false messianism. In obtaining the leading European powers' support, even if qualified, for Zionism, it proved that the road to achieving Zionism's goals of immigration, land purchase, and eventual disposition was a political one. It required the support of the great powers, which could only be achieved by taking

4 The account here is based largely on Weizmann's autobiography, *Trial and Error*, and on the detailed and sometimes critical narrative offered by Jehuda Reinharz in his book *Chaim Weizmann: The Making of a Statesman*. An especially important contribution to understanding the seminal nature of Weizmann's diplomatic work in the period between the Balfour Declaration and the grant of the League of Nations Mandate can be found in Israel Kolatt's article, "Chaim Weizmann's Rise to Leadership," in Isaiah Berlin and Israel Kolatt, *Chaim Weizmann as Leader*, Jerusalem: Hebrew University–Institute of Contemporary Jewry, 1970.

advantage of political opportunities as they presented themselves. In this, Weizmann's rising star was without a doubt a continuation of Herzl's political Zionism.

But the process by which Weizmann consolidated his leadership shows that he developed tools and strategies, both personal and political, that were not available to Herzl. While it is difficult to prove that he consciously analyzed Herzl's inability to translate the impressive web of contacts and meetings he had woven into political achievements, it is clear that he had a different modus operandi. Herzl, perhaps because of his background as a journalist, had a penchant for more or less open diplomacy and was sometimes overly impressed with public statements and the mere fact of meeting with high officials. Weizmann, for his part, perhaps thanks to his background as a *shtetl* Jew with well-honed survival instincts, understood that successful public diplomacy had to be based on a carefully spun web of long-term contacts that would bear fruit, if at all, only years later. As Herzl himself acknowledged, his sense of urgency sometimes impelled him to make dramatic moves – such as his trip to Palestine to meet the German emperor and his attempt at direct negotiations with the sultan – that in the end had no results. Weizmann, by contrast, demonstrated more patience and a pragmatic ability to conduct daily behind-the-scenes activity. In a sense, it was the diplomatic equivalent of the practical settlement policy pursued by the pioneers in Palestine – acre by acre, cow by cow.

That Weizmann lived in distant Manchester, far from the turbulent centers of political power on which Herzl had set his sights and, no less significantly, from the offices of the Zionist movement, made it possible for him to consolidate support and influence, first on the local level and then on the national British scene. In Manchester it was relatively easy to gain access to the editor of the major newspaper, the *Manchester Guardian*, Charles P. Scott. Weizmann's connection with Scott soon turned that elitist newspaper, influential in Liberal circles, into one of the most important disseminators of the Zionist idea in Britain. Weizmann made contact with Balfour and other national leaders in the same way, just as he gained access to Lloyd George through his acquaintance Harry Sacher of the *Guardian*. Weizmann's entry into

British social life taught him the importance of restraint, a trait that Zionist leaders, eager for publicity, did not always display. Self-control, he learned, was the way to make inroads into the English political elite. At the same time, Weizmann took a page from Herzl and rested his arguments for Zionism on a moral and humanitarian foundation and on the Jews' connection to their ancestral land, as well as on Great Britain's political and imperial interests. He was also keenly aware of the need to readjust this blend to match the personality of his interlocutor and the changing conditions created by the fluid political developments of the war and the stormy years that followed.

This is not the place to go into the details of Weizmann's contribution to obtaining the Balfour Declaration, which have been widely discussed in the historical literature. Neither will I here render judgment on the extent to which Weizmann's research on acetone production was a factor in his ability to gain British support for a Jewish homeland in Palestine. There can be no doubt that, at the very least, it enabled him to broaden his connections in Britain's political, industrial, and military leadership. Weizmann's economic independence also enabled him to act independently of the official Zionist institutions, which had in any case been enervated during the war. What is clear is that Weizmann showed up at his political meetings with assets and not just demands in his briefcase. As a consequence, he was able to neutralize opponents in the Zionist movement, mostly people hailing from the German-speaking lands, who remained set in the political orientation that Zionism had displayed prior to the war. Those opponents continued to maintain in 1917, especially following Russia's exit from the war, that Germany would be the victor. On the opposite side, Weizmann's ability to conduct a frank real-political and unsentimental dialogue with British statesmen enabled him to defuse some of the opposition among British Jews to their government's support of Zionism.[5]

The first step that Weizmann took after the Declaration and the

5 Ironically, one of the two British cabinet members who voted against the Declaration was the only Jewish minister, Edwin Samuel Montagu, Secretary of State for India. He was the driving force behind the insertion into its text of the passages

conquest of southern Palestine by the British was the establishment of the Zionist Commission, which went to Palestine while the war was still in progress. In the governmental vacuum created by the end of Ottoman rule and the imposition of British military rule, while the future of the territory remained unclear, the commission created significant facts on the ground. Lacking a mandate from any Zionist body, without any official grant of authority from Britain, and in the face of opposition from the British military administration in Palestine, Weizmann managed to establish a situation in which the general language of the Balfour Declaration was turned into practical measures. The commission was established in January 1918 with the consent of the British government, and its members came exclusively from the Entente powers – Britain, France, and Italy. Its presence in Palestine created a far-reaching and novel reality, in which the Zionists co-opted a de facto role in the administration of Palestine long before any kind of international agreement endeavored to determine its future.

Weizmann arrived in Palestine at the head of the commission in April 1918. Despite his lack of clear official status, he conducted himself as if he represented a political authority. He and his colleagues were received in Jerusalem, Tel Aviv, and the Jewish farming villages with what could only be called royal honors; the ceremonies held for them in different locations were almost official in character. That Weizmann and the other commission members were often accompanied by uniformed British officers (most of them Jewish, such as Edwin Samuel and James de Rothschild) served to enhance the official aura that surrounded them. The speeches made by the commission's members at these receptions made an explicit connection between the body on which they served and the commitments made by Britain in the Balfour Declaration. What had originally been designated as a fact-finding delegation virtually transformed itself into a quasi-administrative body. Without having received sanction from anybody to do so, Weizmann declared the commission to be the official representative of the Yishuv. He set precedents

regarding the rights of non-Jews in Palestine and the status of Jews in other countries.

in his contacts with the British military administration, and, since no other body had authority to speak on behalf of the Yishuv, he assumed powers and made decisions in matters of welfare, education, and immigration absorption. His actions were not always in step with British military policy, but his connections in London enabled him to surmount these obstacles.[6] Weizmann took initiative on the political and diplomatic levels as well. On June 4, he met with Emir Faisal in Transjordan in an attempt to do what he could to moderate the promises that Britain had made to the Arabs and reach an agreement with a prominent Arab leader; and on July 24, at his instigation, he attended a ceremony on Mt. Scopus in which the cornerstone of the Hebrew University was laid. General Allenby, the commander of the British forces in Palestine, was in attendance; he had originally opposed the ceremony, but Weizmann managed to overcome his objections. This, at a time when northern Palestine had yet to be taken by the British; the front was not far away, stretching from Oudja in the east, just north of Jericho in the Jordan Valley, to the Yarkon River, north of Tel Aviv, in the west.

All these actions created facts in the midst of uncertainty, at a time when other Zionist leaders preferred to wait until formal political decisions were made. In this way, Weizmann and the Commission laid the foundations for what would later become the National Council (the Va‘ad Le’umi) and the other elements of Jewish and Zionist self-government during the Mandate period. Precisely because Weizmann was aware of just how problematic the Balfour Declaration was in so many ways (its tenuous status in international law; the opposition of other countries, France among them; and the status it granted the Arabs, to name a few), he understood that realizing its principles required the creation of facts on the ground in Palestine during the interregnum. Without action, the declaration would remain empty verbiage.

Weizmann did not restrict himself to Palestine. He stepped into the international arena, where Palestine’s future would be decided. Here,

6 Film clips from the Zionist Commission’s 1918 sojourn in Palestine can be seen on YouTube: www.youtube.com/watch?feature=player_embedded&v=TVVM9HJNwA (accessed January 24, 2017).

too, Weizmann worked to institutionalize the Balfour Declaration, while at the same time seeking formal recognition of his leadership in the Zionist movement. At the end of 1918, an improvised Zionist Assembly, convened in London and attended only by delegates from the Entente countries, elected Weizmann, along with Sokolow, to head the Zionist delegation to the Versailles peace conference. For the first time, Weizmann had official authorization to speak in the name of the Zionist Organization in diplomatic negotiations. If he had previously been improvising, he now began to consolidate a new type of leadership and political orientation, in keeping with the new positions of the European powers following the victory of Britain and its allies and the defeat of Germany and Turkey.

It was not the Balfour Declaration alone but also the political action that Weizmann pursued between 1917 and 1920 that decided the fate of Palestine. Only in retrospect does it seem obvious and inevitable that Britain should have received a League of Nations mandate for the territory, explicitly charging it with creating a Jewish homeland. Despite the secret Sykes-Picot Agreement between Britain and France, which assigned Palestine to Britain, it was not at all clear at the time that Britain would rule there. No one knew about the Sykes-Picot compact until the Soviets made it public in October 1917. Like other secret agreements signed while hostilities were still in progress, it had no recognized international standing and could easily have been modified in light of the sometimes radically changing military and political circumstances that followed the war.

Weizmann's effort had several dimensions, some on the international stage and others inside the Zionist camp, and these two aspects were intertwined. His first goal was to ensure that the debate over and decisions about Palestine would be conducted separately from the larger discussion of the future of the Levant and Greater Syria. He succeeded in this despite Balfour's departure from the Foreign Office following political changes in Britain in the summer of 1919. He also focused on establishing that Britain would be directly and solely responsible for the fate of Palestine, and that the territory's administration would not be handed over to an international commission. Sokolow, by contrast, toyed with the idea of making France and Italy co-administrators of Palestine. But Weizmann understood that Britain was now an ally of the Zionist movement, while French and Italian support for the Zionist project was illusory, despite Sokolow's meetings with statesmen from those two countries. Unlike Sokolow, Weizmann could tell the difference between a confluence of interests and empty diplomatic talk.

On the other side, Weizmann battled the Zionist maximalists, a sometimes strange coalition of the socialist Po'ale Zion party, Jabotinsky's nationalists, and even Israel Zangwill in one of his incarnations. This group demanded that the international documents granting Britain a mandate over Palestine state explicitly that the goal of the mandate was the establishment of a Jewish state. It was not easy for Weizmann to explain that, at this stage, the term "Jewish national home" could not be interpreted as meaning the establishment of a Jewish state and that no such obligation could be codified in the mandate's charter. There simply were not political and demographic grounds for realizing such a demand under the current circumstances. Furthermore, such a demand, which would disregard the political rights of Palestine's Arabs, would clearly contradict the Wilsonian principle of national self-determination that was a central principle of the peace treaties emerging from the world war (even if that principle was often flouted, as became clear in the treaties of Versailles and Saint Germain). Most of the people demanding the immediate establishment of a Jewish state lacked the political and diplomatic experience that Weizmann had accumulated. Caught up in their dreams, they were inattentive to the international mood and could not accurately assess what could and could not be achieved.

Weizmann's diplomacy bore fruit at the San Remo conference of April 1920, which established the policy of the Entente states regarding territorial and political arrangements in the Middle East following the Ottoman Empire's surrender. The victorious powers confirmed the arrangements envisaged by the Sykes-Picot agreement, while making some changes in it. At the same time, they granted a special status to Palestine, including Transjordan. Britain would rule the territory in accordance with the Balfour Declaration, which for the first time was transformed from a private British document to one ratified by the

Entente powers. This paved the way for incorporating the Declaration into the mandate charter. The Balfour Declaration, that great Zionist achievement, was now a certified component of the postwar international political order.

Three months after the end of the San Remo conference, in July 1920, Weizmann was elected president of the Zionist Organization at the Zionist Conference in London, the first international Zionist gathering since the Eleventh Zionist Congress of 1913. The war still cast a shadow over the London conference: A number of delegates from Germany were unable to obtain visas to enter England. The composition of the delegates was also new. American Zionists took a leading position, reflecting the postwar order in the world and in the Zionist movement. German Zionism lost its hegemony to the movement's English and American wings, while Russian Jewry, under the new Soviet regime, was cut off from the Jewish and Zionist world. The delegates to the conference were not elected, since elections still could not be held; they were appointed in a series of complex and sometimes problematic ad hoc agreements. But the conference gave the movement a new, solid structure, including its new president. Some of the prewar presidents may have been outstanding figures in their own fields, but they lacked political and leadership experience and public stature. Now the movement placed at its head a well-known figure of recognized international standing, with a wealth of experience and achievements.

It was Weizmann's activity between 1917 and 1920, which was prominently reported in the media, that placed him at the head of the Zionist movement. At the time, his work in achieving the Balfour Declaration, which had been accomplished largely behind the scenes, was less known. Rather, it was his labor on behalf of the Zionist cause that placed him at the head of the movement. He understood the importance of acting boldly, even in the absence of any mandate or legitimacy for doing so, but he also grasped the value of moderation and of making clear distinctions between the possible and the pipe-dream. It was this discernment that enabled him to lead the Zionist diplomatic and political effort. At a time of severe crises, of political and administrative chaos, he saw the need to step bravely forward and assume power. That power, and the risks it brought with it, were the foundation on which he was in the end granted formal authority. Tomáš Masaryk of Czechoslovakia and Józef Piłsudski of Poland acted similarly in the whirlwind of World War I. Weizmann's election as president of the Zionist Organization, later ratified, in keeping with the organization's charter, at the Twelfth Zionist Congress, held in 1921 in Carlsbad, marked both the victory of Zionist diplomacy, in deed and not just in word, and the victory of Chaim Weizmann himself on his road to leadership of the Zionist movement.

Shulamit Volkov

Weizmann and His Scientific Colleagues in Germany
Competitors and Standard-Bearers

In his memoirs, Shmaryahu Levin offers an enthusiastic account of crossing into Prussia on the train that took him to the West from Czarist Russia. "The instant I crossed the frontier, and arrived at Eidtkuhnen, a village adjacent to Russian territory, I already felt like a liberated slave," he writes. "... Had I been of a pious turn of mind, I would have recited the prayer of the freed prisoner, and if I had not been ashamed I would have embraced and kissed the Prussian gendarme who now boarded the train."[1]

Perhaps Chaim Weizmann had a similar experience at the same border crossing, but if he did, all memory of it seems to have been obliterated during his initial days in Germany. Like Levin, he traveled to Germany to pursue his studies, in the face of the *numerus clausus*, the quota the Russian authorities had imposed on the number of Jewish students admitted to Russian universities.[2] But while Levin set his sights on the big city, Berlin, Weizmann left Pinsk for Pfungstadt, a small town not far from Darmstadt. He quickly sank into what he experienced as a metaphorical mire.[3] The town was home to a Jewish boarding school,

1 Shmaryahu Levin, *Youth in Revolt* (English transl. by Maurice Samuel), New York: Arno Press, 1975, p. 221.

2 On Weizmann's early years, see Reinharz, *CW*, I.

3 Weizmann describes the experience in his autobiography, *Trial and Error*, pp. 44–49.

which hired Weizmann to teach Hebrew and Yiddish while he pursued his studies at the nearby Technical University of Darmstadt. This work-study plan seemed like a good one at first. He was, above all, eager to begin his courses in chemistry. Of all his teachers in Pinsk, only his chemistry instructor had inspired him, and, despite Pinsk being run-down and poor – "low lying and malarial,"[4] as Weizmann later wrote – and his school day long and grim, he managed to do at least an hour of chemistry each day and to spend two or three days a week in the laboratory. That seems to have been enough to capture his interest and to set him along at least one part of his future path. At the same time, still in Pinsk, Weizmann took an active role in the Zionist movement, beginning the other important and, as it would turn out, the major part of his future life. In other words, his signature combination of science and Zionist political activity was already in evidence in his youth.

Pfungstadt turned out to be a hard slog. His German was weak, his workload heavy, and his studies demanding. "It was an obnoxious place," he wrote of the boarding school.[5] Eight months later he fell ill, apparently with pneumonia, and found himself returning by train from the vaunted freedom of the West to the Russian steppe he so detested. But his second attempt to study in Germany, a year later, was much more pleasant. This time he went to Berlin and enrolled at the Royal Technical College of Charlottenburg. With a monthly allowance from his father and brother-in-law, he was now free of the need to work to pay for his tuition and expenses.[6] Like Levin, Weizmann joined the local organization of Russian-Jewish students, among them some other young men who would later become Zionist leaders, such as Leo Motzkin and Nachman Syrkin. He turned a new page in his life in an entirely new environment. Having mastered German, he was able to engage with the treasures of German culture and observe the special standing of Germany's Jews as active participants within it.

By then, German Jews stood out precisely in those fields that

4 *Ibid.*, p. 35.

5 *Ibid.*, p. 49.

6 *Ibid.*, pp. 50–57.

gradually became the center of Weizmann's academic interests – the exact sciences. Still, while the proportion of Jewish students in Prussian universities was seven times that of the percentage of Jews in the population as a whole, only a handful of Jews held full professorships.[7] In 1910, only twenty-five Jews (who had not converted to Christianity) held appointments as "ordinaries," that is, full professors, in all of Germany's universities, most of them in medical faculties. The notable Jewish presence in the sciences was a product, in part, of the relatively large numbers who served at lower academic ranks and as *Privatdozents*. Most of all, it grew out of the fame of a relatively small number of prominent figures. This had already been true in the nineteenth century, but it became much more salient during the first three decades of the twentieth.

A number of these prominent Jewish scientists, inside and outside the academy, were chemists. At the time, chemistry was a field particularly suited to those seeking to climb the social ladder. Medicine was the classic profession for achieving higher status, but by the end of the nineteenth century, chemistry, particularly organic chemistry, had begun to serve as a launching pad for students seeking rapid advancement, both Jews and non-Jews. Naturally, the academic path proved suitable for only a few, but by the end of the century some six hundred academically-trained chemists worked in full-time positions in German industrial firms; in Britain, at the same time, there were no more than forty. Today even the German numbers seem very low, but it was clear at the time that this was only the beginning. Chemistry looked like a promising field where all practitioners would be judged solely by their professional accomplishments and where Jews, too, would be free, at least to some extent, of restrictions based on national and religious affiliation.

Weizmann writes that he was "greatly overawed" by his fellow-students in the Russian-Jewish association. "I was only a student of chemistry; they were students of philosophy, history, economics, law, and other 'higher' things," he relates. This might not have been an entirely

7 Figures can be found in Shulamit Volkov, *The Magic Circle: Jews, Antisemites, and Other Germans*, Tel Aviv: Am Oved, 5762 (2001/02), pp. 209–223 (Hebrew).

true picture, but it reflects Weizmann's values at the time and his precise professional goals. "In later years I understood that even deeper motives impelled me in those days to attend strictly to the question of my personal equipment for the life struggle," he writes. "For the time being it was enough for me to make up my mind that I was going to achieve independence."[8] Perhaps without his intending it, the fact that he was a chemist may have later given him special standing in the Zionist movement. Science provided him with financial means, and in some periods even with a measure of affluence, but most importantly it made him economically self-sufficient over the course of his life. It was a common practice for Zionist activists to threaten to resign their positions in the movement; Weizmann could actually carry out his threat because, as he often said, he could always return to his laboratory.

That proved to be both an advantage and a disadvantage. Early in his career, Weizmann was not personally acquainted with famous scientists who devoted themselves entirely to science. He would only later meet Albert Einstein, James Franck, Richard Willstätter, and Fritz Haber. But during his time in Germany he did occasionally encounter people who were wholly devoted to science, figures for whom science was a calling. From the outset, this was the model he measured himself against. But he knew that he was not like them. Even then he was a man torn between two worlds. He would never become a scholar who could devote himself solely to the ideals of pure science.

This duality, an inherent part of the way he lived, had in fact vexed Weizmann as early as his Pinsk days, but the tension between the two poles intensified considerably in Germany and always remained a sore point, especially during later visits to Germany and in his contacts with German colleagues. The motif of "the deep division in my life, or perhaps I should say its organic duality," as he put it, appears repeatedly in his memoirs. He spoke of a ceaseless "double pull, towards science on the one hand, towards a public life in the Zionist movement on the other."[9] And even if it seems unquestionable that his real passion was

8 *Trial and Error*, pp. 54–55.

9 *Ibid.*, p. 76.

politics, his tone and choice of words indicates that he experienced it as a profound inner conflict. In Russia, he explains, his focus on science was seen as "cowardly and selfish,"[10] whereas in Germany, he gives us to understand, it was considered a supreme ideal. Since he adopted the latter view, he continually viewed himself as not having lived up to the ideal to which he aspired, and perhaps even as having betrayed it.

Weizmann completed his doctorate successfully and submitted it to the University of Freiburg, in the French-speaking part of Switzerland. After a short period at the highly-regarded University of Geneva, he moved to England, where he found himself in an entirely different academic environment. At the University of Manchester he mainly encountered scientists of a practical bent. These were hard-working researchers seeking solutions to practical and concrete problems, people with less pronounced theoretical ambitions, with whom he felt he could successfully compete. The most prominent of them was clearly Ernest Rutherford, "as an experimenter ... a genius,"[11] Weizmann wrote. He was the polar opposite of an intellectual far removed from the issues of this world, like the "mandarins" who had won his admiration in Germany.

Although in England Weizmann could more easily have reduced his public work so as to devote himself more fully to science, he never actually did so. On the contrary, it did not take long before he discovered that the British scientific ideal made it easier for him to focus on Zionist affairs while still continuing his research activities. Furthermore, in England he could leverage his successes in one field in order to achieve success in the other. The tension that had so rankled him in Germany seemed to have abated. As a scientist, he worked to achieve important, even vital, extra-scientific goals, such as arming the British fleet and army, and he did so along with other scientists for whom this combination was a matter of course. And on top of addressing Britain's needs in World War I, he also promoted the Zionist cause. He was now no

10 *Ibid.*, p. 91.

11 *Ibid.*, p. 154. Weizmann writes of his early activity in England in *Trial and Error*, pp. 142–155, and of Rutherford on pp. 153–155.

different from his colleagues, both those who worked or competed with him in England and those working in France, Belgium, and Switzerland, with whom he maintained close professional ties.

Only in 1921, by which time Weizmann was president of the World Zionist Organization, did he have an opportunity to renew his connections with the German scientific community, confronting once again its values and modes of operation. The bridge to that community was this time no other than Albert Einstein. Since Weizmann's major project at that time was the establishment of the Hebrew University in Jerusalem, he decided to send Einstein a telegram and invite the celebrated physicist to accompany him on a trip to the United States. One of the purposes of the trip was to raise money for the university project. The telegram was brought to Einstein by Kurt Blumenfeld, the leader of Germany's Zionists. Unsurprisingly, Einstein demurred, arguing that it would not be proper to take advantage of his scientific fame to mobilize the masses for other than purely scientific goals. But Blumenfeld did not take no for an answer. If Einstein was serious about his Zionism, Blumenfeld told him, he was obligated to join Weizmann on the trip. To Blumenfeld's astonishment, Einstein reversed himself and said that he had no choice but to accept the mission.[12] "Einstein's decision reflected a major transformation in his life," writes his biographer, Walter Isaacson. Up to that point, Einstein had devoted himself exclusively to science, the only exceptions being personal and private matters. He now made a real leap "not of faith but of commitment," asserts Isaacson.[13]

Not everyone accepted that change lightly. Felix Frankfurter and Julian Mack, leaders of the Zionist movement in the United States, wrote to Weizmann in alarm to ask him to cancel his planned trip with Einstein, or as they put it, "the tragedy of his visit." It would be seen, they

12 The episode is recounted in full in Blumenfeld's memoirs. See Kurt Blumenfeld, *Erlebte Judenfrage: Ein Vierteljahrhundert deutscher Zionismus*, Stuttgart: Deutsche Verlags-Anstalt, 1962, pp. 129–130. It also appears in biographies of Einstein, e.g., Ronald W. Clark, *Einstein: The Life and Times*, New York–Cleveland: World Publishing Company, 1971, pp. 465–466; Walter Isaacson, *Einstein: His Life and Universe*, New York: Simon & Schuster, 2007, pp. 290–291.

13 Isaacson, *Einstein* (above, note 12), p. 291.

warned, as a commercialization of science, and it was bound to rouse the ire of scientists at American universities.[14] But their warning came too late: Einstein had already set out. In any case, most of his visit ended up being taken up with scientific lectures, and his Zionist activity did not in any way detract from the honors with which he was received. In Germany, Fritz Haber felt particularly uncomfortable about Einstein's trip to the country that had been Germany's enemy during the recent world war. He wrote to Einstein and asked him to reconsider. Such a trip, he maintained, would be taken as evidence of German Jews' disloyalty to their country. "You will certainly sacrifice the narrow basis upon which the existence of professors and students of the Jewish faith at German universities rests," he added. While Haber did not explicitly argue that a Zionist advocacy trip by Einstein would sully the ideal of pure science, he was furious at Einstein for committing to the trip with Weizmann while canceling his participation in the Solvay Conference in Brussels, the first such conference to be held after World War I, symbolizing the return of German scientists to the international scientific fold. Einstein rejected Haber's appeal out of hand. On March 21, 1921, he set sail to New York with Weizmann.[15]

For many years after this trip, Weizmann and Einstein remained in almost continuous contact. The correspondence between them demonstrates a genuine relationship between the two. Their disagreements

14 Quoted from their letter of March 19, 1921, to Weizmann, in Ze'ev Rosenkranz, *Einstein before Israel: Zionist Icon or Iconoclast?*, Princeton–Oxford: Princeton University Press, 2011, p. 105.

15 The exchange of letters between Haber and Einstein is quoted at length in Dietrich Stoltzenberg, *Fritz Haber: Chemiker, Nobelpreisträger, Deutscher, Jude*, Weinheim: Wiley-VCH, 1998, pp. 291–393. On the relations between the two, see Fritz Stern, "Freunde im Widerspruch. Haber und Einstein," in Rudolf Vierhaus & Bernhard vom Brocke (eds.), *Forschung im Spannungsfeld von Politik und Gesellschaft: Geschichte und Struktur der Kaiser-Wilhelm-/Max-Planck-Gesellschaft*, Stuttgart: Deutsche Verlags-Anstalt, 1990, pp. 516–551. Weizmann recounted the trip to the United States in detail in his memoir: *Trial and Error*, pp. 330–342. He also described the conflicts within the Zionist movement that ended up making the trip tense and problematic. [See Section I in the chapter "Unpublished Passages" in this volume. – Ed.]

over the academic administration of the Hebrew University in its early years are well known,[16] as are their political differences – among other issues, regarding attitudes and policies toward the Arabs. Perhaps it was the fact that they could discuss science between them, and not only politics, that preserved their relationship despite these differences. While their letters offer no evidence of this, they apparently had such discussions on the way to the United States. At their first press conference after coming ashore, Isaacson relates, Weizmann was asked if he understood the Theory of Relativity. He replied: "During the crossing, Einstein explained his theory to me every day, and by the time we arrived I was fully convinced that he really understands it."[17] Isaacson calls this a "delightful reply," but it was more than that. It showed Weizmann explicitly acknowledging that Einstein the scientist was a cut above Weizmann the scientist. Indeed, throughout the trip Weizmann found himself in the great man's shadow. The crowds cheered Einstein and welcomed him like a star, in Isaacson's account. When Weizmann spoke, they listened politely.[18]

While Einstein agreed to accompany Weizmann on this trip, he afterward turned down nearly every request Weizmann made of him. The difference between them was clear: Einstein was, above all, a scientist, devoted entirely to his scientific work. Weizmann was a politician. In some circles, for the most part British ones, his scientific work was highly regarded, but he was always divided between two worlds. In the final analysis, his devotion to Zionism proved stronger.

The same tension colored his relations with two other important German scientists, Fritz Haber and Richard Willstätter. Unlike Einstein, Willstätter was in a very real sense Weizmann's colleague. Both were organic chemists, and they understood each other perfectly on all scientific matters. They could also collaborate. Willstätter, unlike Einstein but like most Jewish scientists in Germany, was *not* a Zionist. He was a Jew, true, but of a very specific variety. Some branches of his family

16 [On this debate, see the article by Issachar Unna in this volume. – Ed.]

17 Isaacson (above, note 12), p. 292.

18 *Ibid.*, p. 294.

had utterly assimilated, but when his admired mentor, Adolf von Baeyer (also of Jewish extraction), suggested to the young, talented, and ambitious young chemist that he convert to Christianity, Willstätter refused. While he had never lived any sort of Jewish life, conversion seemed to him, as it did to many other Jews at the time, a dishonorable deed. The second act of this drama is no less fascinating. It took place in the summer of 1924, when Willstätter, who had received the Nobel Prize in Chemistry in 1915 and was one of the world's best known and most important organic chemists, surprised everyone by resigning his post at the University of Munich. As he related in his autobiography, in doing so he gave up the residence the university had supplied him with, and he "never again set foot" in the magnificently-equipped laboratory that had been provided for him there.[19] His resignation was a ringing act of defiance against the university, which had just refused to appoint the chemist and mineralogist Victor Goldschmidt to a vacant chair in his field. Willstätter maintained that the only reason Goldschmidt was not appointed was that he was a Jew. Willstätter's totally unexpected resignation demonstrated that this esteemed scientist was after all a Jew of conscience and above all a person of principle – a fitting interlocutor for Weizmann.

The two had first met prior to World War I. Blumenfeld relates in his memoirs that, after the war, Weizmann sent him to Willstätter to recruit him to the Zionist cause, but Willstätter refused to have any contact with Weizmann. "He is responsible for the deaths of German soldiers," Willstätter maintained.[20] They established a real acquaintance only at the end of 1931 or the beginning of 1932, after Weizmann's resignation from the presidency of the World Zionist Organization and his return to the scientific work he had neglected for thirteen years. At this point he asked the assistance of Willstätter, who was still living in Munich but traveled extensively. The two met in London and became friends and

19 Richard Willstätter, *Aus meinem Leben: Von Arbeit, Muße und Freunden*, ed. Arthur Stoll, Weinheim: Verlag Chemie, 1949, pp. 341–351.

20 Blumenfeld (above, note 12), p. 168.

collaborators. Even more importantly for historians and biographers, they began to exchange letters.

The Weizmann Archive in Rehovot holds the entire correspondence. A small portion of it is devoted to purely scientific matters, which I am not qualified to evaluate. But most of it is personal – warm and sometimes even moving. Willstätter always made a point of addressing Weizmann with deference. There is nothing of the sharp and somewhat arrogant tone typical of Einstein's letters to Weizmann. He was always willing to offer assistance on scientific matters, was excited when they were scheduled to meet and carefully planned for such encounters. Moreover, he was always prepared to travel – to Ostend in Belgium, to Zurich, to Paris, and even to London – wherever Weizmann wanted to meet him, he wrote, adding: "By the way, I am willing to come to meet you even if there is no need for it in connection with work."[21] He took an interest in Weizmann's family and never forgot to send his warm regards to Weizmann's wife, Vera. Moreover, Willstätter admired Weizmann for precisely what Weizmann himself sometimes saw as a betrayal of science, namely, his intensive Zionist activity. Weizmann complained that he was called on again and again to take part in politics and that this caused delays in his research. Willstätter empathized and offered encouragement. For example, in June 1933 he wrote:

> I am very sorry that, despite the decision to free yourself of organizational and political obligations and to concentrate on your quiet and productive scientific work, in which you have already had several successes because of your exceptional talents in the field of practical chemistry, new tasks have nevertheless been imposed on you, taking advantage of your willingness to help and to make sacrifices.[22]

The combination of public activity and scientific work, which

21 Willstätter to Weizmann, CWA, 30-1634 (May 8, 1933; appended to this article). See also, e.g., his letters of November 24, 1932 and February 14, 1933 – CWA, 15-1600 and 9-1618.

22 Willstätter to Weizmann, CWA, 28-1646 (June 16, 1933).

Weizmann viewed as an insult to the ideal of total devotion to research, won Willstätter's admiration: "The fact that, in the midst of all the anxieties of our time, you devote so much energy [to your laboratory] is a real wonder."[23] Willstätter made every effort not to place any further burdens on his friend. He continually recommended that Weizmann rest and take vacations, and he reiterated his promise to go anywhere at any time to meet his new friend. He also offered an explanation of the tension felt by Weizmann, juxtaposing the fastidious and slow work of the laboratory and the pressures that were an inevitable part of any political activity. He marveled at Weizmann's "sublime achievements," at his great "life work," adding that "in contrast, my own work seems worthless."[24] There can be no doubt that Weizmann liked to hear this. Willsätter, the scientific authority he held in the highest esteem, endorsed the path he had chosen and accepted his renunciation of a purely scientific career.

Willstätter seems to have been shaken in some way by his contact with Weizmann. He stressed how enriching it had been for him.[25] Their meeting, he wrote to his friend in 1933, a few months after Hitler's assumption of power in Germany, "was very meaningful for me, and it provided me with new hope and new horizons, at a time when my entire life seemed negative and depressing."[26] But those promising new horizons were not enough to detach Willstätter from his surroundings and bring him to Palestine, as Weizmann had planned. Weizmann offered his German friend the post of director of the Sieff Institute, only to be turned down. Willstätter visited Palestine one time, in 1934, for the institute's opening celebration, where he gave the keynote address. Weizmann received him as a guest of honor both privately with his family and in political and scientific circles. But Willstätter then

23 *Ibid.*

24 Willstätter to Weizmann, CWA, 19-1844 (September 17, 1935; appended to this article), 17-916A (June 19, 1936), and 10-2320 (September 9, 1941; also appended here).

25 Willstätter to Weizmann, CWA, 23-1689 (December 1, 1933).

26 Willstätter to Weizmann, CWA, 28-1646 (June 16, 1933).

returned to Europe and apparently never seriously considered the possibility of leaving. Nevertheless, something changed in his approach to Judaism. Upon returning from his visit to Palestine, Willstätter unexpectedly agreed to write an introduction to a large volume on the Jewish contribution to German culture that the Jüdischer Verlag, a Berlin Jewish publishing house, planned to put together. It is hard to believe that he would have undertaken such a thing before.[27] He even informed Weizmann that he was trying to learn Hebrew, and in a subsequent letter he related that he had read Josephus Flavius's two histories and was, for the first time in his life, reading the Old Testament.[28] He admired the accomplishments of the Sieff Institute and did all he could to encourage Weizmann in his endeavors to expand it.

But by this point there was not much else he could do. The letters portray a person whose self-confidence had suffered a heavy blow as a result of events in Germany. His health was tenuous and his spirits low. His contact with the energetic Weizmann, brimming over with plans, had a salutary effect on him. While Einstein responded negatively or even categorically rejected nearly every proposal Weizmann made – for example, that of recruiting young Jewish scientists in Germany to move to Palestine[29] – Willstätter saw himself as a friend, duty-bound to offer encouragement and support. "Of all my experiences in recent years, the biggest impression came from my meeting with you, and the brief path we have walked together on several occasions," he wrote.[30]

They met for what was apparently the last time in October 1939 in Zurich, immediately after World War II broke out. Weizmann had gone to France at the behest of the British government for political and scientific talks.[31] One of his assignments was to discover whether there

27 In the end, the book was published only after World War II: Siegmund Kaznelson (ed.), *Juden im deutschen Kulturbereich: Ein Sammelwerk*, Berlin: Jüdischer Verlag, 1962.

28 Willstätter to Weizmann, CWA, 8-1724 (February 16, 1934) and 34-1736 (April 19, 1934).

29 Einstein to Weizmann, CWA, 21-1762A (July 26, 1934).

30 Willstätter to Weizmann, CWA, 21-258 (June 16, 1939).

31 I rely here on Menachem Kedem, *Chaim Weizmann in the Second World War:*

was any truth to reports that the Germans possessed secret chemical weapons. It was with that subject in mind that Weizmann met with Willstätter, whom he described to his British operators as a person who had possession of all the required information, and who was "reliable, precise, and thorough." Willstätter indeed briefed him on the state of German research. He claimed that the Germans had the capability to use only chemical agents of the type they had had in World War I. He then, according to Weizmann's report to the British Foreign Office, offered his views on the larger situation. While he believed that Germany would ultimately be defeated, it would not happen with a single determined counter-offensive, and not before the full significance of what he called "totalitarian war" became salient. This was a form of war that the Allies did not comprehend, he said. Germany would prove willing to sacrifice its own population in order to achieve military successes.

This love story has a strange epilogue. In his memoirs, Weizmann wrote of his empathy for Willstätter's difficulties and offered a glowing portrait of his personality and of their friendship. Yet, oddly, in his own memoirs, which he completed in December 1941, shortly before his death, Willstätter mentions Weizmann only once, in passing, in connection with his trip to Rehovot in 1934. That is all. There is nothing about their friendship, nothing about his admiration for the Zionist leader. Perhaps he expected more active assistance from Weizmann after he lost his university position and had to move to Switzerland. Or perhaps the story of this friendship seemed to him detrimental to the image of himself that he sought to paint in his book. Did he, at the end of his life, seek to represent himself only as a German scientist? Was that why he said nothing about Weizmann, thinking that it would undermine the picture of his unwavering identification with Germany? Perhaps it was precisely this that had made his correspondence with Weizmann so emotional? Did the entire affair eventually embarrass him? There are no clear answers.

Here are the intimations of tragedy in Weizmann's relations with his

Weizmann and Zionist Policy in the Years 1939–1945, Tel Aviv: Ma'ariv, 1983, pp. 26–31 (Hebrew).

German Jewish scientific colleagues. But the tragedy is salient in the story of Weizmann's acquaintance with Fritz Haber. Haber gained fame as a result of his discoveries and inventions during World War I.[32] The most important of these were related to the production of synthetic ammonia, but Haber was also professionally and personally involved in the production of poison gasses for military use, and for the practical application of his inventions during the war. Born Jewish, Haber had converted to Christianity and was a fervent German patriot. Still, he was so highly regarded as a scientist that he was awarded the Nobel Prize in Chemistry in 1918, even before the end of the war. Weizmann had no personal ties to him. Given that Haber had opposed Einstein's trip to the United States with Weizmann, and that he stood for everything Weizmann hated about German Jewry – repudiation of all things Jewish, utter assimilation, and what Weizmann saw as servility – there was no basis for real friendship. "I therefore had no desire to meet Haber," Weizmann wrote in his memoirs.[33]

Nevertheless, even before the Nazis came to power, Weizmann visited Haber's laboratory in Berlin and established a kind of initial, though somewhat hesitant connection with him.[34] Soon thereafter, the Nazis brought about Haber's resignation from his post as head of the Kaiser Wilhelm Institute of Chemistry in Berlin and drove him into exile. During long months of going from hotel to hotel all over Europe, Haber had trouble deciding what to do. Surprisingly, he turned to Weizmann,

32 See Daniel Charles, *Master Mind: The Rise and Fall of Fritz Haber, the Nobel Laureate Who Launched the Age of Chemical Warfare*, Cape May, NJ: Ecco, 2005. And see in particular Fritz Stern, "Fritz Haber: The Scientist in Power and in Exile," in idem, *Dreams and Delusions: The Drama of German History*, New York: Alfred A. Knopf, 1987, pp. 51–76. See also Stern's lecture at the centenary of the Fritz Haber Institute in Berlin on October 28, 2011, "Fritz Haber: Flawed Person and Country." I am grateful to Joshua Jortner for bringing this text to my attention and supplying me with a copy.

33 *Trial and Error*, p. 435.

34 [See Section II in the chapter "Unpublished Passages" and Illustration 31 in this volume. – Ed.]

who immediately came to his aid.[35] "I went to him at once, and found him broken, muddled, moving about in a mental and moral vacuum," Weizmann related. Haber was apparently still grasping at the hope that he would somehow be permitted to leave Germany honorably, with proper emigration papers and while maintaining at least some of his considerable property. Weizmann did all he could to help, involving even his contacts at the British Foreign Office. But, as was to be expected, it came to naught. In the end, Weizmann suggested that he move to Palestine and join the Hebrew University, and although Haber had received other offers, he immediately consented. He then traveled to Zermatt, at the foot of the Matterhorn, to meet Weizmann again, and was now given a more precise offer, including a well-equipped laboratory and private housing. That evening, as Weizmann spoke of his political plans and of his doubts concerning these plans, Haber interrupted and urged him to attend the forthcoming Eighteenth Zionist Congress in Prague. Weizmann, Haber insisted, had to carry on his grand enterprise, which "will stand, a shining monument, in the long history of our people."[36] Against all odds, Weizmann seems to have succeeded in lighting a spark of Jewish identification in this man, always so distant, just as he had with Willstätter.

"I was never in my life so Jewish as now!" Haber wrote to Einstein.[37] But he addressed that comment to the wrong person. Unlike Weizmann, Einstein was unforgiving. "I am happy that your love for the blond beast has cooled some," the physicist replied. "Who could have believed that my dear Haber would appear before me as a representative of the Jewish, even Zionist, cause?"[38] Weizmann, by contrast, took Haber's intentions very seriously. He was aware, of course, that Haber was conducting parallel negotiations with the University of Cambridge, but he continued to lobby Haber to choose Jerusalem, to inform him of

35 *Trial and Error*, pp. 436–437.

36 *Ibid.*, p. 437.

37 Quoted in Charles, *Master Mind* (above, note 32), p. 229.

38 Einstein to Haber, AEA, 12-387 (August 9, 1933). [Other passages from this letter appear in the article by Issachar Unna in this volume. – Ed.]

affairs in Palestine, and to promise him great things. He was solicitous and encouraging.[39]

But it was too late. Haber, who had suffered from severe angina pectoris throughout this period, died in Basel in January 1934. Weizmann cabled Willstätter: "I mourn with you the death of a dear friend and colleague, a servant of science and humanity."[40]

Many would take exception to this description of Haber. But at that moment Weizmann did not see fit to condemn him. His heart went out to such people, his Jewish scientific colleagues from Germany. They had been in error, he believed, but the retribution they endured was far greater than anyone could have expected. In the end, none of the great scientists whom Weizmann tried to recruit ended up coming to Palestine. Yet their scientific dedication and values took root in Israel. That should not be taken for granted. They always had to be fought for, as they still do.

39 See, e.g., Weizmann to Haber, CWA, 14-665A (October 2, 1933), 2A-1671 (October 20, 1933), and 2-1689 (December 1, 1933). Weizmann wrote in his memoirs that Haber died on his way to Palestine; see *Trial and Error*, p. 438. Haber's letters to Weizmann indicate, however, that he canceled his plans to go to Palestine some three weeks before his death, and that in any case he had intended only a visit and not a move. See Fritz Haber, "Letters to Chaim Weizmann," *Leo Baeck Institute Year Book*, 8 (1963), pp. 103, 105, 109; and Stoltzenberg, *Fritz Haber* (above, note 15), pp. 612, 615, 621–628. I am grateful to Benjamin Kedar for enlightening me on Haber's change of heart.

40 Telegram, Weizmann to Willstätter, CWA, 3-1717A (January 31, 1934).

Appendix: From Weizmann's Correspondence with Richard Willstätter[41]

1. Willstätter to Weizmann, May 8, 1933. CWA 30-1634

Professor Richard Willstätter
München 27
Möhlstr. 29

den 8. Mai 1933

Sehr verehrter, lieber Herr Kollege Weizmann:

Danke schön für Ihre liebenswürdige Begrüssung vom 6. ds. und bin froh, dass Sie wohl zurück vom Osten sind, wo Sie leider diesmal trotz des Abstands von europaeischen Geschehnissen nicht soviel Behagen und Ruhe gefunden zu haben scheinen.

Unser Wiedersehen wird nur davon abhängen, wann es Ihnen nach dem Stand der Arbeiten u. dgl. erwünscht ist. Mir passt von heute bis Anfang August jede Zeit, und ich komme mit grösstem Vergnügen zum Rendez-vous, nach London, Paris, Zürich, Ostende. Übrigens komme ich ebenso gern, Sie zu treffen, wenn die Arbeiten es nicht nötig machen.

Herzlich grüssend Ihr aufrichtig ergebener
Richard Willstätter

41 These handwritten letters were transcribed by B.Z. Kedar.

2. Willstätter to Weizmann, September 17, 1935. CWA 19-1844

Professor Richard Willstätter
München 27
Möhlstr. 29

den 17. Sept. 1935

Sehr verehrter, lieber Freund:

Herzlichen Dank für Ihren liebenswürdigen Brief vom 10. ds.; ich bin immer froh, einen von Ihnen diktierten englischen Brief zu bekommen, und ich erkenne Miss May mit Vergnügen schon von weitem.

Sie empfanden es als schweren Kampf, den Entschluss zur Annahme der Präsidentschaft zu fassen. Aber aus grösserer Entfernung gesehen, war der Gang der Ereignisse lang vorbereitet, zwangsläufig, unausweichlich. Es war störend und peinlich und unerträglich geworden, dass es in entscheidenden Phasen der palästinensischen Entwicklung eine formale und eine wirkliche Präsidentschaft gab. Andererseits war doch das Kaleidoskop der menschlichen, wirtschaftlichen und politischen Angelegenheiten, die unaufhörlich Ihr Denken und Ihr Eingreifen beanspruchten, in störende Konkurrenz getreten mit dem ruhigen Ablauf der wissenschaftlichen Arbeiten. Sie waren wieder mit tiefer Neigung Biochemiker geworden. Aber während in den Aufgaben der Zeit und ihren Unruhen ein Semester Rehovoth schon zu einem langen und einem bewegten Zeitraum wird, vermögen wir in den Laboratorien erst in Jahren und Jahren geduldiger Vertiefung die unscheinbaren Handhaben unerwarteter einzelner Beobachtungen zu einem Werk zu gestalten. Und das Werk will nur gedeihen, wenn wir es, so lang es unfertig, ständig überschätzen, ohne es kritisch zu vergleichen mit den grossen Erkenntnissen, die vor uns zu stande kamen und die neben uns reifen. Die Konkurrenz der so rasch strömenden ausserwissenschaftlichen Interessen und der so langsam fliessenden chemischen Forschung im Bereich Einzelner musste bei Ihnen Beunruhigung und Verstimmung erzeugen, wie sie sich bei jenem kurzen Spaziergang nahe Ihrem Hause am Abend des 30. Juni anzudeuten schien. Künftig werden Sie die befriedigende

Lösung solcher Konflikte darin finden, dass die Grösse und Fülle Ihrer Aufgaben, wovon Sie nur für die bedeutenderen selbst eintreten wollen, Ihnen das langsame Fliessen der wissenschaftlichen Arbeit erträglich machen wird, und dass Sie sich in Ihren Verpflichtungen die Beschränkung auferlegen werden, die nötig ist, um trotz aller Abhaltungen doch Ihr Institut von Rehovoth zu gedeihlichem Zusammenarbeiten und die Forschung zu richtigen Ergenbnissen zu führen.

Für die gütige Erinnerung an Ihre in London angedeutete Einladung nach Rehovoth bin ich sehr dankbar. Aber ich fürchte, im kommenden Winter und Frühjahr nicht dafür frei zu sein. Während meine Freunde mir anraten wollen, von hier wegzugehen, halte ich mit Freudigkeit aus und zwar bei einer Arbeit, die ich im letzten Herbst begonnen hatte und die ich in diesen Tagen nochmals von vorn beginne. Es ist ein unerwartetes wahres Glück für mich, dass ich noch einmal in meinem Leben meine Kraft an etwas wagen darf, was schwierig ist, und was mir als nicht unwichtig erscheint. Ich wäre gerne bereit mit einem 'noli turbare circulos meos' unterzugehen, aber es wird nicht dazu kommen. So lang diese Arbeit im Gang sein wird und in der Zeit, die schliesslich für ihre schriftliche Erstaltung nötig sein wird, habe ich keine Freiheit.

Bitte, bestellen Sie Ihrer verehrten Gemahlin und Herrn und Frau Simon Marks meine verbindlichen Grüsse und seien Sie selbst herrzlich gegrüsst von

Ihrem aufrichtig ergebenen
Richard Wilstätter

3. Willstätter to Weizmann, September 2, 1941. CWA 10-2320

Locarno-Muralto, 2. Sept. 1941

Hochverehrter, lieber Freund:

Diese Zeilen schreibe ich, noch ohne zu wissen, wohin, oder auch nur in welchen Erdteil sie zu senden sind, aber ich versuche noch ein Mal, durch Ihre Züricher Freunde eine Adresse zu erfahren. Seit Ihrem

Besuche in Locarno, der mir so lieb und wichtig war, dachte ich immer mit tiefem Anteilnehmen und mit Sorge an Sie selbst und an Ihre liebe Familie; ich hoffe Sie und Ihre verehrte Gemahlin wohlbehalten und besonders auch Ihren tapferen jüngeren Sohn, der sich bei der R.A.F. so rühmlich bewährt und ausgezeichnet hat. Hoffentlich ist es auch der Familie Blumenfeld verhältnismässig gut gegangen!

Eine Nachricht empfing ich seit Ihrem Besuche nicht. Sie beabsichtigten damals, mir Ihre späteren Patentschriften und einen Bericht über die Versuche mit Blutserum zu senden.

Es ist eine grosse Erleichterung, wahrzunehmen, wie die Verlagerung des Krieges an die Ostfront die drohende, unmittelbare Gefahr von Palästina abgelenkt hat, wenigstens für geraume Zeit. Die grossen Distanzen im Osten lassen für die in engen Räumen erprobten besonderen Methoden der deutschen Heere geringe Chancen. Man kann dort siegen, aber siegen bedeutet nicht gewinnen. Mit den Distanzen sind auch die Zeitspannen gewachsen. Ihre Absicht, im nächsten Jahre in die Schweiz zu kommen, wovon ich erfuhr, ist sehr willkommen, aber auch sehr unsicher. Ich fürchte, es wird länger dauern.

In Eremitaggio ist es ruhiger als noch im vorigen Jahre. Ich konnte das Buch meiner Lebenserinnerungen zu Ende schreiben und ein wenig weiter arbeiten, wenig. Inzwischen ist meine Lage finanziell ungünstig geworden. Durch die Deportierungen aus Südwestdeutschland wurden nämlich Verwandte und Freunde betroffen; nach einer Leidenszeit, die sie im Lager durchmachten, habe ich ihnen geholfen, bis zu meinem letzten Franken und weit darüber hinaus. Für sieben Personen bestritt ich die Atlantic passages und für eine Anzahl Personen sorge ich weiter. Zur selben Zeit wurde mir der Rest meiner Einkünfte und meines Besitzes in Deutschland genommen, und, was mehr bedeutet, auch Alles an industriellen Rechten. So lebe ich und helfe Anderen nur durch einen schmalen Licenzvertrag [*sic*] mit Prof. Stell. Im übrigen sehe ich mich, wie zu erwarten war, von allen Freunden und Schülern verlassen.

Ich muss an die in Aussicht gestellte Entwicklung Ihrer Proteolysearbeiten denken, wofür Ihr freundschaftliches Vertrauen mich Anfang 1933 zur Mitarbeit gewann und wofür Sie die Güte hatten, mir ein Anrecht einzuräumen. Es ist mir unbekannt, ob Verwertungen, die Sie für unmittelbar bevorstehend hielten, und wofür die gegenwärtigen Verhältnisse besonders geeignet wären, erfolgten, oder ob Sie diese Angelegenheiten aufgeschoben haben. Es gibt heute natürlich wichtigere Arbeiten. Immerhin ist es mir schmerzlich, von dem Fortgang nichts zu wissen und wieder ein Jahr und ein Jahr und wohl die Zeit meines Lebens die Erwartung zu vertagen. Sie wissen, dass ich mich mit unbegrenztem Vertrauen mit Ihnen verbunden habe. Auch heute noch suche ich an einer letzten Hoffnung festzuhalten, das, was Sie so freundlich in Aussicht stellten, werde nicht zu spät für mich anfangen, in Erfüllung zu gehen. Im besonderen handelt es sich um die von Ihnen in Aussicht gestellte Verwertung in U.S.A. Ich trat dafür ein, diese müsste von einer a priori-Bareinzahlung abhängig gemacht werden. Unter den heutigen Umständen ist ein solcher Erfolg für mich notwendig geworden, heute.

Schreibe ich Ihnen heute zu offen und zu sachlich? Ich hoffe, nicht. Ihnen gegenüber habe ich mich, wie niemals einer anderen Person gegenüber, stets gehemmt gefühlt in jeder Aussprache über praktische Dinge. Das war vor allem die Folge meiner tiefen Verehrung für Sie, in der ich mir immer vorkam, wie wenn ich der weit Jüngere wäre. Und es war ferner eine Folge meiner respektvollen Empfindungen für Ihr grosses Lebenswerk, dem gegenüber mir die andersartigen Ziele meines Lebens gering erscheinen mussten. Aber es ist wohl eben die Grösse Ihrer Lebensaufgabe, die Ihnen daneben eine praktische Auswirkung von Laboratoriumsarbeiten und die Berücksichtigung damit zusammenhängender Interessen zu geringfügig erscheinen lassen wird.

Hoffend, diese Zeilen werden Sie in bester Gesundheit und unverminderter Zuversicht erreichen, bin ich mit meinen herzlichen Grüssen

Ihr aufrichtig ergebener
Richard Willstätter

Benny Morris

Weizmann and the Arabs

Let me begin with a curious fact. In September 1907, Chaim Weizmann visited Palestine for the first time. He spent a fortnight in the country, touring its "length and breadth," as he later put it.[1] He visited the mixed Arab–Jewish towns of Jerusalem and Haifa and the Arab town of Nazareth, lodged for a few days in Jaffa, a mostly Arab town, and on his travels passed through or by strings of Arab villages. At the time, the country was inhabited by some 600,000 Arabs and 60–70,000 Jews. During his visit, Weizmann wrote and sent off a handful of letters – to his wife, Vera, who remained in Manchester, and to Zionist industrialist Johann Kremenetzky, in Vienna. After his return to England, he sent a number of letters describing his visit to acquaintances. In none of these letters did Weizmann mention an "Arab" or "Arabs."[2] It was as if he had not encountered or seen any, as if the land was without Arabs, as if only Jews populated its hills and valleys, streets and alleyways. It was only a few weeks later, in a speech in Manchester, that Weizmann at last referred to Palestine's Arabs, speaking of their "primitive attachment to the land" and that they might some day begin to feel that they had a moral right to the country.[3]

Acknowledgments: I would like to thank Merav Segal and Keren Raz in the Chaim Weizmann Archives and Orna Zeltzer in the Weizmann Institute Archives for their efficient help in finding the documents I required relating to Chaim Weizmann and his work in the Institute.

1 *Trial and Error*, p. 162.

2 *CWL*, V, nos. 55–67, pp. 53–70.

3 Norman Rose, *Chaim Weizmann: A Biography*, New York: Viking, 1986, p. 106.

By contrast, David Ben-Gurion, who would eventually emerge as Zionism's foremost leader, noted the Arabs' presence when he first came to Palestine, at about the same time. After arriving on a Russian steamship in September 1906, he wrote home of the Arabs he encountered on deck: "They made a very good impression upon us. ... They are nearly all good-hearted, and are easily befriended. One might say they are like big children." He must have felt the Arabs' presence in the country even before he set foot on the soil of the Promised Land; after all, he was carried by a brawny Arab from the steamship onto a skiff and from the skiff to the shore, as was the custom in Jaffa's harbor in those days, since it couldn't host large ships.

Weizmann must have reached shore in the same manner. And in Jaffa, Weizmann must have seen and met Arabs – the very same Arabs that Ben-Gurion had described the year before:

> Jaffa is not pretty. As in any Oriental city, the streets are narrow and winding. An awful dust hangs over the marketplace. … I saw an uglier exile in Jaffa than in Plonsk [his hometown]. … It was full of Arab shopkeepers who sat about smoking their nargilas (water-pipes). Among them were a few Jews. It made a bad impression upon me.

Ben-Gurion was repelled by the sight and after a mere seven hours fled Jaffa for the Jewish settlements to the east.[4]

Perhaps Weizmann's reaction – in not seeing or hearing the natives or at least in not noting their presence in his letters to Europe – was typical. Around the *fin de siècle*, European settlers and travelers in countries in what is today called the Third World often didn't see or count the indigenous people. It was the heyday of Western imperialism, and the visitor traveled to and between concentrations of Europeans. The masses of natives who served the Europeans were part of the background, an element of the scenery that drew or merited little interest or attention.

4 Shabtai Teveth, *Ben-Gurion and the Palestinian Arabs: From Peace to War*, New York: Oxford University Press, 1985, pp. 6–8.

Weizmann in this respect was a typical liberal European of his generation. He was a democrat with a well-developed sense of social justice, but he did not take issue with Western imperialism and colonialism. Indeed, in his writings he was wont to refer to the Jewish settlers in Palestine as "colonists," to the settlements as "colonies," and to the whole Zionist enterprise as "colonization." In part, this was a matter of linguistics. The early Zionists called their settlements *moshavot* (literally, colonies). But it also reflected the norm of European colonial settlement in other parts of the globe – in the Far East, South Africa and Algeria. Zionism differed from the typical European colonialism, in which imperial mother countries sent their sons to settle distant Third World countries, exploit their natural and human resources, and establish sovereign outposts of strategic worth. The Zionists, by contrast, had come to settle their ancient homeland and, to be sure, to become once again its sovereign masters – not to grab an alien tract of land on behalf of this or that empire or to exploit its resources. In Weizmann's eyes, the Jews were returnees to their homeland, not invaders in the service of an imperial power. But the Zionist settlers were indeed Europeans, and Palestine was at the time inhabited by an overwhelming native (Arab) population. Weizmann didn't see them – until a rival and antagonistic Arab national movement emerged a few years later.

But over the decades, Weizmann's attitude toward the Arabs gradually changed. In the beginning, they were merely "natives." The Zionists, he felt, had to treat them fairly – much as the British ethos demanded fair treatment of native populations around the empire. Weizmann believed in the Arab's right to live in Palestine as he and his forefathers had inhabited the land for generations.[5] But he did not believe that the Arabs had "national" rights, and he did not view them as a collective with an identity separate from that of the surrounding Middle Eastern Arabs – those of Transjordan, Syria, and Lebanon.

In the course of the 1920s and 1930s, as the Palestine Arab national movement emerged, Weizmann came to recognize the separate "peoplehood" of the country's Arabs and their national rights. He reached

5 Report on Weizmann's speech, *Haaretz*, December 19, 1919, p. 2.

the conclusion that in Palestine there was a struggle between two legitimate national movements (though he always, at least in public, stressed what he saw as the destructive nature of the Palestine Arab national movement). Partly as a result of this recognition of the existence of a rival national movement, Weizmann in the end posited the idea of partition and the establishment of an Arab state side by side with a Jewish state. And after the establishment of the State of Israel, Weizmann adopted an egalitarian attitude toward its Arab minority, arguing that they were citizens with equal rights and had to be dealt with fairly.

THE NATURE OF THE ARABS

Like many Europeans who reached maturity at the end of the nineteenth century, Weizmann harbored a "racial" view of peoples and countries. Germans were seen as different from Italians, Jews as different from Germans, and Jews as different from Arabs. The differences between peoples and races may have stemmed from different histories, cultural singularities, or societal variants, rather than from genetic or biological factors, but they were a given. Weizmann view of the Arabs certainly underpinned his attitude toward them and the conflict with them.

In January 1941, following the Palestinian Arab revolt and the anti-Jewish atrocities that accompanied it, and in light of the support expressed by many Arabs for an Axis victory in the war, Weizmann gave vent to his views. Speaking with Ivan Maiskii, the Soviet ambassador in London, he said: "The Arab is often called the son of the desert. It would be truer to call him the father of the desert. His laziness and primitivism turn a flourishing garden into a desert."[6] There is no reason to doubt the

6 So Maiskii recorded Weizmann's comment in his report: "Meeting: I.M. Maiskii – Ch. Weizmann" (London, February 3, 1941), *Documents on Israeli–Soviet Relations 1941–1953*, Part 1: *1941–May 1949*, London: Frank Cass, 2000, p. 5. Weizmann gave a similar report of the conversation to the Jewish Agency Executive, though he attributed the saying to Major C.S. Jarvis, the onetime British governor of Sinai: "Meeting of the Jewish Agency Executive (London, 30 January 1941)," *ibid.*, p. 2. The dates are those of the respective reports, not of the conversation itself.

accuracy of Maiskii's report of the conversation; Weizmann wrote similar things in earlier years. "The chances are that if the Negev is given to the Arabs, [the region] will remain a desert forever," he wrote to an acquaintance in England in 1937.[7] The year before, he had written that the Arabs of Palestine were "an element destructive [of the land]; we [the Jews] build."[8]

Not that, for Weizmann, the desert didn't have a positive side. Like most people, Weizmann, of course, preferred a green and pleasant land to an arid one. But, like many British officers, officials, and travelers, he appears to have seen in the "classic" desert Arab an embodiment of the noble savage. In 1918, after his first meeting, in Transjordan, with Prince Faisal ibn Hussein, who commanded the Arab forces in the Great Arab Revolt of 1916–1918 against the Ottoman Empire, Weizmann made a point of contrasting the Hijazi emir with the notables he had met in Levantine Palestine. In a letter to Vera, he wrote of Faisal:

> [He is] the first true national Arab that I have met. He is a leader! He is intelligent and very honest, and astonishingly handsome. … He is contemptuous of the Arabs of the Land of Israel and does not even consider them Arabs.[9]

Weizmann characterized the notables of Palestine, who would lead the Palestine Arab national movement, as "dishonest, uneducated, avaricious, and unpatriotic."[10]

7 Weizmann to Geoffrey Mander, *CWL*, XVIII, no. 142, p. 158.

8 Weizmann to Walter Baer, *CWL*, XVII, no. 248, p. 268.

9 Weizmann (Tel Aviv-Jaffa) to Vera Weizmann (London), June 17, 1918, in *Chaim Weizmann, the First President: Selected Letters and Speeches*, ed. Louise Fischer, Jerusalem: Israel State Archives, 1995, p. 143. For decades, Weizmann continued to write glowingly of Faisal, even after the latter reneged on the agreements the two had reached in 1918–1919. In his autobiography, composed in the 1940s, Weizmann wrote of their "lifelong friendship" (*Trial and Error*, p. 294). Faisal died in 1933.

10 Mark Tessler, *A History of the Israeli–Palestinian Conflict*, Bloomington: Indiana University Press, 1994, p. 153.

It appears that Weizmann first met Arabs on his way to Palestine in 1907. He spent several days in Alexandria, a city with an Arab majority and a substantial European minority. He found the city "interesting" but also "extremely repulsive":

> A mixture of European "civilization" and Eastern and African barbarism. … All the prerequisites of a cultural life are lacking. Debauchery, licentiousness, passion, cupidity, greed for profit by plunder, murder, rape – all this is to be seen in the faces of the people. This is the threshold of Palestine, and I fear that Egypt, which can exercise a tremendous influence on the development of Palestine, may in fact exert a fatal one.[11]

A few days later, while in quarantine in Beirut, Weizmann wrote that "all kinds of savages are milling about" the area.[12]

Weizmann's long sojourn in Palestine in spring–summer 1918 did little to alter his attitude toward the Arabs. In May 1918 he spoke of Arab "loafers and vagrants" near the Wailing Wall,[13] and to Balfour he wrote of "the treacherous nature of the Arab":

> [They might yet] stab the [British] Army in the back. The Arab … screams as often as he can and blackmails as much as he can … The Arab official is a "roué" and therefore has a great advantage over the fair and clean-minded British official, who is not conversant with the subtleties and subterfuges of the Oriental mind.[14]

11 Weizmann (Alexandria) to Vera Weizmann (Manchester), *CWL*, V, no. 52 (August 29, 1907), p. 50. In fact, Egypt exercised little influence on the development of Palestine during the following decades. Since the 1970s–1980s, via the Hamas, the Muslim Brotherhood's branch in Palestine, Egypt has certainly exerted an indirect influence upon developments in Palestine/Israel.

12 Weizmann (Beirut) to Vera Weizmann (Manchester), *CWL*, V, no. 53 (September 2, 1907), p. 51.

13 Weizmann to William Ormsby-Gore, *CWL*, VIII, no. 185 (May 1, 1918), p. 177.

14 Weizmann to Balfour, *CWL*, VIII, no. 208 (May 30, 1918), pp. 198 and 201.

Decades later, in his autobiography, Weizmann somewhat modulated his assessment, but the negative thrust remained in place:

> The Arab is a very subtle debater and controversialist [*sic*] – much more so than the average educated European – and until one has acquired the technique one is at a great disadvantage. In particular, the Arab has an immense talent for expressing views diametrically opposed to yours with such exquisite and roundabout politeness that you believe him to be in complete agreement with you, and ready to join hands with you at once. Conversations and negotiations with Arabs are not unlike chasing a mirage in the desert: Full of promise and good to look at, but likely to lead you to death by thirst.[15]

Weizmann's view of the Arab national movements, including that of Palestine's Arabs, was no doubt colored by his assessment of the Arab mentality, though it was also influenced by Arab behavior. In January 1919 he wrote that the "[Palestine Arab] native" would be unfit "for self-government for a very, very long time to come."[16] A decade later, Weizmann wrote that "Arab nationalism" in "Palestine, Syria, or Egypt" has "one thing in common" – "xenophobia."[17] He dismissed the leaders of the nascent Palestine Arab national movement, with Musa Kazim al-Husseini at their head, as "fifth-rate people"[18] and "trash."[19] In his memoirs, he would describe the Palestinian leaders during the 1936–1939 revolt as "a set of unscrupulous Levantine politicians."[20]

In the aftermath of the 1929 Wailing Wall riots, Weizmann dismissed any possibility of fruitful dialogue with the leaders of the Palestine

15 *Trial and Error*, p. 271.

16 Weizmann to General Arthur Wigram Money, *CWL*, IX, no. 112 (January 26, 1919), p. 105.

17 Weizmann to Manka Spiegel (Graz), *CWL*, XIII, no. 364 (March 18, 1928), p. 408.

18 Weizmann to Abraham Tulin (New York), *CWL*, X, no. 240 (August 10, 1921), p. 247.

19 Weizmann to Vera Weizmann (Carlsbad), *CWL*, X, no. 243 (August 14, 1921), p. 251.

20 *Trial and Error*, p. 486.

Arab national movement. They were not only intransigent; they were wily and deceptive. To Robert Weltsch, a leading German Jewish supporter of bi-nationalism in Palestine, he wrote:

> The Arab mind is not ripe at all for any negotiations. … They are not producing arguments, but tricks. They are not statesmen but low class politicians with a mentality of Oriental carpetdealers. In your memorandum to us you admit the primitiveness of the Arabs but, at the same time, you urge us to deal with them. Their primitiveness, of which you speak, is a terrible difficulty, and almost insuperable. Unless the Arabs are brought to a different frame of mind I don't see any possibility of successful negotiation.[21]

His views on Arab nationalism only hardened during the Arab Revolt of 1936–1939. In 1937, speaking to the Royal Commission headed by Lord Peel, which was investigating the causes of the revolt and ways to resolve the Palestine problem, he argued:

> Arab nationalism is, if I may say so … a crude imitation of the material side of European nationalism – very crude. They take the external forms; they take … the machine guns and not the moral content, not the renaissance which is usually bound up with it … there is no spiritual side to it. In Palestine, there is not a single picture, not a single song, not a single book, by an Arab nationalist. Their love of the country has not found expression in literature or in any moral achievement, but in machine guns and in the very successful employment of dynamite which has been acquired illicitly. … They are a people who because of lack of culture, because of lack of stability, because of lack of bigger outlook, discount things very rapidly. They are a nervous race, they are very easily depressed, and they are equally easily elated.[22]

21 Weizmann to Weltsch (Berlin), *CWL*, XIV, no. 110 (November 25, 1929), p. 109.

22 Quoted in Joseph Heller, "Weizmann, Jabotinsky and the Arab Question: The Peel Affair," *Jerusalem Quarterly* (Winter 1983), p. 113.

In 1938, Weizmann wrote to Malcolm MacDonald, the British Colonial Secretary, whom he believed at the time to be pro-Zionist: "Arab nationalism is totalitarian in nature, shallow, aggressive and arrogant." He linked this to Muslim religion and culture:

> Islam is a totalitarian creed, and the much-vaunted Arab tolerance, if it exists at all, endures only so long as the minorities agree to remain under-dogs. … No wonder that Arab nationalism, in its teaching and methods, should be modeling itself on the pattern of [German] National Socialism.[23]

Indeed, a year later Weizmann went one better, asserting that Arab nationalism was, at least in one respect, worse than Nazism. Arab nationalism, he wrote,

> reveals unmistakably its character as a crude primitive imitation of German and Italian Fascism from which it derives its inspiration. It is even lower than that [*sic*] in its quality because it lacks completely the constructive attempts at the betterment of the masses noticeable in the former. The *effendis* continue to exploit the ignorance and religious fanaticism of the poor *fellaheen* whose faces they grind.[24]

Weizmann and Arab–Zionist Relations before 1948

Already before World War I, Weizmann became aware of Arab nationalist claims to Palestine. A separate Palestinian Arab national movement would emerge only in the 1920s, but the Syrian Arab national movement, with its epicenters in Beirut, Damascus, and Aleppo, was already claiming Palestine as part of its patrimony. In early 1913, Weizmann wrote to Vera: "There is alarming news from Syria about the Arab national movement. … They consider Palestine their own," and

23 Weizmann to MacDonald, *CWL*, XVIII, no. 367 (July 12, 1938), p. 425.

24 Weizmann to Felix Frankfurter (Oxford), *CWL*, XIX, no. 113 (July 1, 1939), p. 138.

they were already waging a propaganda campaign against land sales to Jews, using, in Weizmann's phrase, "semi anti-Semitic" language. He concluded: "We soon [will] face a serious enemy."[25]

In the course of World War I, the relatively uninformed Weizmann may have contemplated negotiating with "the leading Arabs from Palestine," as he told C.P. Scott, the editor of the *Manchester Guardian*.[26] But immediately after arriving in Palestine in April 1918, at the head of the Zionist Commission, Weizmann formed a poor impression of the notables with whom he came in contact. Among others, he met with Musa Kazim al-Husseini and D.G. Salameh, respectively the mayor and deputy mayor of Jerusalem, and with Kamal Bey al-Husseini, the town's mufti. He may also have met Muhammad Haj Amin al-Husseini, who in the 1930s and 1940s would head the Palestine Arab national movement.[27] The local notables may have sounded moderate and reassuring when speaking in the presence of British officials, but, according to Weizmann, at heart they believed that the Balfour Declaration heralded a policy designed "to oust them from the country."[28]

Weizmann's own placatory reassurances proved unavailing. He came to appreciate the Arabs' fears of dispossession, or at least of subordination to the Zionist settlers. This was brought home to him by two Arab speakers at a public event he attended. Their nationalism, Weizmann complained, was "intransigent and aggressive," and they believed that Jews were robbers bent on enslaving and ruining the Arabs.[29] Weizmann understood that "the Arab problem" was, and would remain, "the central problem" of Zionism.[30] He may have believed at the time, and often said, that Jewish immigration and settlement need not encroach on the Arabs or be at their expense, and that there was room enough in

25 Reinharz, *CW*, I, pp. 385–386.

26 Weizmann to Scott, *CWL*, VII, no. 321 (March 20, 1917), pp. 344–345.

27 Reinharz, *CW*, II, pp. 248–253. The report that Weizmann met Haj Amin may have been untrue, as Reinharz suggests.

28 Weizmann to Louis Brandeis, *CWL*, VIII, no. 175 (April 25, 1919), p. 159.

29 Weizmann to Ormsby-Gore, *CWL*, VIII, no. 161 (April 16, 1918), pp. 128–130.

30 Rose, *Chaim Weizmann* (above, note 3), p. 195, quoting Weizmann in June 1918.

Palestine for both communities.[31] But this did nothing to dispel Arab fears.

Weizmann, disappointed with the local notables – who were encouraged in their anti-Zionism by members of the British military administration[32] – began to look to an agreement, under British auspices, between the Zionist movement and Arab leaders outside Palestine, especially the Hashemites, mainly meaning Faisal, who had led the Hijazi rebel army to Damascus, and his brother Abdullah, who was appointed by the British as Emir of Transjordan in 1921–1922. Weizmann believed, or at least hoped, that if the World War I victors granted independence and sovereignty to the Arabs in the whole expanse from Beirut to the Persian Gulf and southward as far as Yemen, the region's leaders would agree to the award of sovereignty to the Jews over the narrow coastal sliver of Palestine. It seemed only fair. Faisal came close to saying this in their agreement of January 1919,[33] and Weizmann hoped that it might be possible to separate the regional leaders from the Palestine Arab notability and national movement. In 1931, Weizmann put it this way:

> We say to the Arabs that they must distinguish between their "national home" and their rights in Palestine. The "national home" of the Arabs is in Damascus and Baghdad, and in the religious sense, perhaps, in Mecca and Medina, those great centres of Arab life, in which a mighty Arab culture flourished. It flourished in Baghdad and Damascus, not in Jerusalem.[34]

A few years later, incidentally, the Peel Commission would say more or

31 *Ibid.*, p. 198, quoting Weizmann in 1918, and p. 239, quoting Weizmann in 1924.

32 Weizmann to Vera Weizmann, *CWL*, VIII, no. 151 (April 6, 1918), p. 119, and no. 200 (May 20, 1918), pp. 190–191.

33 See the text of the agreement in George Antonius, *The Arab Awakening: The Story of the Arab National Movement*, Philadelphia: Lippincott, 1939, pp. 437–439.

34 Michael Assaf, "On the Arab Question," in Meyer Weisgal (ed.), *Chaim Weizmann: Statesman, Scientist, Builder of the Jewish Commonwealth*, New York: Dial Press, 1944, p. 58.

less the same thing: In the glorious Arab empires of the Middle Ages, Jerusalem had always been an unproductive backwater.[35]

But Arab nationalism had a dynamic of its own, and the rise of a separate Palestine Arab national movement resulted in resistance to and absolute rejection of such a compromise, with the Palestinian leaders, as it turned out, feeding Arab intransigence elsewhere and Arab leaders elsewhere feeding and reinforcing Palestinian resolve. Weizmann and other Zionist leaders might continue seeking and hoping, through the 1920s, 1930s, and 1940s, to break through the vise of pan-Arab rejectionism, but the promise they offered – that Zionism would lead to the enrichment and development of Palestine and the region as a whole – failed to overcome the basic Arab hostility to Zionism's central goal: the establishment of Jewish sovereignty in Palestine. The economic benefit that Zionism undoubtedly brought to Palestine failed to outweigh the local Arabs' growing fears that the Zionists ultimately meant to uproot and dispossess them.

In 1918–1920, in light of his extremely positive meetings with Faisal, Weizmann believed that it would be possible for the Zionists to reach an agreement with "the Arabs" through the Hashemites, with the assistance of their joint patron, the British. He reported that at their December 1918 meeting in London, Faisal had said that "the advent of the Jews into Palestine was for the good of the country" and that "complete harmony" prevailed between the Arab national movement and Zionism.[36] A few days later, on January 3, 1919, the two men signed their famous agreement. In it, Faisal explicitly supported the Balfour Declaration and endorsed massive Jewish immigration to Palestine, while Weizmann promised to support the aims of extra-Palestinian Arab nationalism, politically and economically.[37]

35 *Palestine Royal Commission Report*, July 1937, p. 5.

36 Memorandum by Weizmann on the meeting in the Carlton Hotel, London, December 11, 1918, attached to Weizmann to Eyre Crowe, *CWL*, IX, no. 70 (December 16, 1918), pp. 70–71.

37 Articles 3, 4 and 7 of the Faisal–Weizmann agreement; Antonius, *Arab Awakening* (above, note 33), pp. 438–439.

But Faisal almost immediately reneged on the agreement, and in the following year, in July 1920, he was driven out of Damascus by the French. (In 1921, the British crowned him king of the newly fashioned Kingdom of Iraq, where he ruled until his death thirteen years later.) But in truth, even in the months when he ruled in Damascus, from October 1918 until July 1920, Faisal never governed the Arab expanses east of Egypt or even all the territory of present-day Syria. His area of effective control was limited to the Damascus region. Faisal never became the potent Arab leader he aspired to be. Indeed, viewed in the wider context, the Arab national movements of the first decades of the twentieth century failed to produce a leader who could represent and unite them, and the Zionist movement was unable to find an Arab leader with whom it could reach a binding agreement about the future of Palestine. Weizmann understood this, and fairly quickly after signing his agreement with Faisal, he recognized that it was of little significance, if not actually a dead letter. Or, as he put it in a letter to Herbert Samuel, soon to be named the first British high commissioner in Palestine (1920–1925):

> The main difficulty [is] in the fact that there is no responsible head of the Arab Nation with whom one can deal here. They are broken up in factions, each led by one or two agitators who give vent to their feelings in abusing us grossly, threatening us and insulting us sometimes very vulgarly.[38]

Zionist–Arab peacemaking efforts were plagued by the diversity of power centers in the national movements in the surrounding Arab states and by factionalism among Palestine's Arabs. In the mid-1930s, however, the Palestine Arab national movement unified around the figure of the mufti of Jerusalem, Haj Amin al-Husseini, who overcame the "Opposition" led by the Nashashibi clan. The struggle between the two had often pushed the factions into assuming and enunciating

38 Weizmann (Jaffa) to Samuel (London), *CWL*, IX, no. 240 (November 22–23, 1919), p. 256.

strong anti-Zionist positions as a means of winning popular support. The Husseini victory over the Opposition in 1936–1937 put paid to any thought of an Arab–Zionist compromise, down to the Palestinian Nakba (catastrophe) that was the obverse side of Jewish statehood in 1948. Husseini rejected every effort at conciliation and, indeed, spoke explicitly of denying citizenship rights to most of the country's Jews. In his testimony before the Peel Commission in 1937, he went even further, hinting at the expulsion of all or many of the country's 400,000 Jewish inhabitants, and the commission pointedly referred to the Iraqi massacres of Assyrian Christians a few years before in connection with what could be expected of the Husseini-led Palestinians should they assume control of the country.[39] Husseini rejected the idea of partitioning Palestine into two states, along with the idea of a unitary bi-national state, advocated by radicals in the Jewish community as represented by Brit Shalom (the peace alliance) and Agudat Ihud (the unity association).

Weizmann, for his part, continuously emphasized that the Jews would bring development and progress to the country and that the condition of Palestine's Arabs would be bettered as a result. In a 1949 interview, he quoted T.E. Lawrence ("Lawrence of Arabia") as having said that "only through cooperation with the Jews would the Arabs grow."[40] In 1943, Weizmann appealed to Palestine's Arabs:

> If you and Jews [*sic*] will cooperate, you will build up, for your mutual benefit, and for the benefit of the world, those tremendous areas which today are a desert and a reproach to humanity. Such an upbuilding, indeed, is a life and death necessity for you. ... You and the Jews can make this part of the world safe for yourselves and for us [in the face of predatory imperial powers].[41]

39 *Palestine Royal Commission Report* (above, note 35), p. 141.

40 "Interview given by Dr. Weizmann to the Representative of the 'Times' on 11.1.49," CWA, 3-2883.

41 "No Patchwork Solution, A Word to the Arabs – to the United Nations – and to Certain Jews, Message from Chaim Weizmann," *The New Palestine*, May 7, 1943, CWA, 2441A-11.

And he was not thinking only of Palestine's Arabs. In 1949, he maintained that the newborn State of Israel "might also play an important part as a pioneer of scientific research and development in the Middle East [as a whole]."[42] Specifically, he believed that as a result of scientific advances, mainly in hydrology and agriculture, the Negev could be developed, and the country could absorb three–four million Jews. So, at least, he told President Harry Truman.[43]

Neither the leaders of the Arab states nor the Arabs of Palestine were tempted or persuaded by Weizmann's vision of development and progress accompanying the Zionist enterprise. The Arabs continued to believe that the Zionists' aim was to uproot them. In the first years after the British conquest of Palestine there were moments when Weizmann was optimistic regarding a possible Arab–Zionist agreement – or at least, he broadcast such optimism to his British interlocutors. For example, in September 1919 Weizmann passed on to Lord Curzon, a member of Lloyd George's Cabinet, what he regarded as "evidence of the rapprochement which is in progress" between the Jews and the Arabs in Palestine.[44] In retrospect, however, Weizmann appears gradually to have sunk into pessimism, which only burgeoned with time. Already in 1920, he expressed strong doubts about the possibility of a settlement. He pointed to a number of factors. "Sharifian and Turkish agents," he wrote, were stirring the pot of animosity. Moreover, the nationalism preached in elite circles in Palestine, such as in the Muslim–Christian associations, was "not free from a strong religious element, and the nationalist teaching assumes often the form of hostility against the infidel [Jew]," he wrote. The Arabs were being told that the Zionists were entering the country to despoil them "of their land and property" while the *effendi* class had turned against the Jews out of fear of losing their privileged status. Italian agents and French propaganda, both driven by Catholicism, were also fostering anti-Zionist feeling. To these he added

42 "Interview … 11.1.49" (above, note 40).

43 Weizmann to Truman, CWA, 1-12618 (December 12, 1945).

44 Weizmann to Curzon, *CWL*, IX, no. 224 (September 22, 1919), p. 224.

the contribution of the British administrators, who radiated anti-Zionism, so that the Arabs quickly saw "which way the wind was blowing."[45]

Weizmann tried to explain the British officers' and officials' hostility toward Zionism. He wrote:

> [The official], especially in the lower ranks – prefers the native to the Jew not out of any reason of unfairness or anti-Semitism, but simply because the native is a much simpler proposition than the Jew. ... The Jew is bristling with problems, ... and sometimes bewilders the administrator. The administrator is [simply] used to handl[ing] natives.[46]

Nonetheless, it is likely that Weizmann also believed that, among some officials, their hostility toward Zionism was motivated by anti-Semitism.

Weizmann's evolving pessimism was doubtless reinforced by events on the ground. In April 1920 he was in Jerusalem during the so-called Nabi Musa riots, when six Jews were murdered in the city, dozens were injured, and several women apparently were raped. He wrote to Vera: "We went through something absolutely frightening and agonizing – a pogrom in Jerusalem, with all the accompaniments and 'joys' of a pogrom." He blamed the British military administration, which, while it may not have "actively organize[d] it ... undoubtedly played a passive role." He described the British officials as "wolves and jackals" who had immediately disarmed and arrested the Jewish self-defense groups. Otherwise, he opined, these armed Jews would have overcome the Arabs on day one.[47]

Weizmann, like other Zionist leaders, continued to want, and to seek, a deal with the Arabs. Perhaps he saw peace as a long-term existential necessity for Zionism's survival; perhaps he saw in it a supreme value, as, indeed, it figures in parts of the Bible; perhaps he thought the movement's leaders must continuously, outwardly, display an interest

45 Weizmann to Curzon, *ibid.*, no. 280 (February 2, 1920), pp. 296–300.
46 Weizmann to Money, *ibid.*, no. 112 (January 26, 1919), p. 106.
47 Weizmann (Italy) to Vera Weizmann, *ibid.*, no. 306 (April 19, 1920), p. 336.

in peace, even while not believing that it was realistic or attainable, at least in the short term. Be that as it may, his will for peace was not a product of whim or of a desire for immediate political gain. From here derived his tendency to attribute the absence of progress toward Arab–Zionist peace to technical reasons, or to external factors like the British bureaucracy.

Weizmann wrote to Wyndham Deedes, a senior pro-Zionist Mandate official: "I am not afraid of negotiations with the Arabs, but we have never been given a clear chance to come together and negotiate frankly"[48] – as if to say that the British were responsible for the absence of such direct contacts, which, if enabled, might have led to a deal. Weizmann repeatedly wrote, and perhaps believed, that if the British had resolutely favored Jewish–Arab peace and unequivocally supported the establishment of a sovereign Jewish state, a peace settlement would have been achieved. British tergiversation or lack of resolution, he believed, was partly responsible for Arab rejectionism and animosity.[49]

But in truth, Weizmann, from the 1920s onward, understood that the problem ran deeper: The goals of the two national movements in Palestine were diametrically opposed, the gap between them was unbridgeable, and it was irrelevant how many specific, practical obstacles were removed and how many rounds of negotiation were entered into.

In 1930–1931, in the wake of the 1929 Wailing Wall riots (which the Arabs called a "revolt"), Weizmann – like Ben-Gurion – briefly toyed with the idea of a bi-national state, at least as a transitional measure until a final, permanent settlement could be reached. Perhaps a joint Jewish–Arab government could be established, he suggested.[50] But Weizmann always insisted on political parity between the two communities as the basis for a bi-national arrangement, despite the acute disproportion between the sizes of the two communities (at the start

48 Weizmann to Deedes (Jerusalem), *CWL*, X, no. 315 (December 13, 1921), p. 329.

49 Weizmann to Felix Warburg (New York), *CWL*, XIV, no. 63 (September 24, 1929), pp. 49–52, and Weizmann to Warburg, *ibid.*, no. 89 (November 13, 1929), pp. 72–82.

50 Rose, *Chaim Weizmann* (above, note 3), pp. 288–291.

of the 1930s there were 200,000 Jews and close to a million Arabs in the country). Weizmann also insisted that Jewish immigration (*'aliya*) must not be restricted, in the present or in the future. In proposing the bi-national idea, Weizmann hoped that eventually, a Jewish majority would emerge, and then the bi-national state would be replaced by a state encompassing all of Palestine in which there would be a Jewish majority. In any event, in 1932 he abandoned the bi-national idea, and his hesitant "adoption" of it in 1930–1931 can be seen as no more than a passing political tactic.[51]

The 1920s passed without any progress toward resolving the conflict. In the mid-1930s, Weizmann, like other Zionist leaders, reached the conclusion that the Zionist movement could not achieve sovereignty over the whole Land of Israel. This was brought home by the Palestine Arab revolt that erupted in spring 1936, forcefully demonstrating the presence and power of a rival – Palestinian Arab – national movement. During the revolt, which lasted until spring 1939, the British also came to appreciate the movement's strength and gradually abandoned their commitment to Zionism. (The gathering war clouds in Europe and the Far East and the triple threat from Germany, Italy, and Japan were a major factor in this abandonment.)

In 1936–1937, Weizmann and Ben-Gurion were persuaded that there was no escaping a partition of Palestine. Not that Weizmann came to support the Palestinian cause; on the contrary, during those years he always stressed, at least in public, the negative nature of the Palestinian Arab national movement:

> On one side, the forces of destruction, the forces of the desert, have risen, and on the other stand firm the forces of civilization and building. It is the old war of the desert against civilization.[52]

51 Weizmann to Weltsch, *CWL*, XIV, no. 110 (November 25, 1929), pp. 107–110, and Weizmann to James Marshall (New York), *ibid.*, no. 187 (January 17, 1930), pp. 205–211.

52 *Davar*, April 23, 1936, p. 1.

But he came grudgingly to recognize the existence of the rival movement and its goals – independence and sovereignty – and saw them as on a par with, if not on an identical standing with, the goals of Zionism.

In early 1937, Weizmann met with members of the Peel Commission and either helped convert them, or was converted by them – it is unclear which – to the principle of partition as the necessary, unavoidable solution to the Palestine conflict. In February he was quoted as saying: "The Jews would be fools not to accept [partition], even if it [i.e., the proposed Jewish state] were the size of a tablecloth."[53] At the back and front of Weizmann's mind were not only the Arab Revolt and the gradual British acknowledgement of Arab rights, or at least British fear of the Arabs, but also the growing distress and threatened destruction of European Jewry. Weizmann was convinced of the need to establish, immediately, a Jewish state, even a minute one, that could provide a safe haven for many of Europe's Jews.

At the same time, Weizmann, like Ben-Gurion, may have thought that a mini-state in part of Palestine might not be the last word in the matter; as he hinted to Stephen Wise: "The problems of expansion and extension" of the mini-Jewish state could be left to "future generations."[54] Nonetheless, from that point on, Weizmann, in public and in political debate, would be a firm and unwavering advocate of a settlement based on partition.

But in 1938–1939, the British Government, under Neville Chamberlain, executed a *volte face* in its Palestine policy, finally and definitively rejecting the idea of partition and the establishment of a Jewish state – which it had formally and publicly endorsed in July 1937, after the publication of the Peel Commission report. In the White Paper of May 1939 (the "MacDonald" White Paper), London declared its support for the grant of independence to Palestine within ten years, the country to then be governed by the majority of its inhabitants – with an Arab majority assured by the stipulated limitation of Jewish immigration to Palestine during the following five years to 75,000, with all Jewish immigration

53 Rose, *Chaim Weizmann* (above, note 3), p. 320.

54 *Ibid.*, p. 323.

thereafter contingent on Arab agreement (which all understood would not be forthcoming). Taken together, these provisions nixed the idea of partition and Jewish statehood and assured the emergence of an Arab majority state with a Jewish minority.

Weizmann, aware of the gist of the impending White Paper, wrote in April 1939 to the pro-Zionist Conservative MP Leopold Amery that the new policy would reduce the Jewish "National Home," as proposed and supported by the British Government in the Balfour Declaration of 1917, "to a ghetto," and would hand over "Jewish Palestine to the Mufti and his gang."[55] To Prime Minister Chamberlain, the shocked Weizmann cabled that the new policy would augur the "destruction of Jewish hopes and surrender [the] Jewish community [of] Palestine to rule [by the] junta responsible for [the] terrorist campaign." The Jews would fight Britain rather than bow to such a policy, he implied.[56]

During the years of World War II, as European Jewry was being murdered by the Nazis and their many helpers, the British government carried out the enunciated policy of severely curtailing Jewish immigration to Palestine (and, simultaneously, of severely restricting further Jewish land purchases in the country). But the war's outbreak in September 1939 also in effect put on hold the proposed grant of independence to an Arab-majority Palestine. The war and the Holocaust only reinforced Weizmann's determination to quickly establish a Jewish state, in any part of Palestine. His letters and speeches of 1939–1941 reveal a sense of impending catastrophe and an overriding urgency to achieve Jewish statehood to obtain a safe haven for Europe's Jews.

In 1941–1942 the safe haven came under serious threat when Germany's Afrika Korps, under General Erwin Rommel, crossed the Egyptian border and appeared poised to conquer the Nile Valley. Weizmann later recalled a correspondent who arrived in the U.S. from Palestine and related that "the Arabs were already preparing the division of the spoils [i.e., Jewish property]. Some of them were going about the streets of Tel

55 Weizmann to Amery, *CWL*, XIX, no. 46 (April 25, 1939), p. 44.

56 Weizmann to Chamberlain, *ibid.*, no. 41 (April 18, 1939), p. 37.

Aviv and the colonies [the Jewish settlements] marking up the houses they expected to take over."[57]

But events turned out otherwise. Rommel was beaten and thrown out of North Africa, and Palestinian Jewry was left unmolested. But Germany, before crashing, murdered most of Europe's Jews. From then on, the Holocaust, which Weizmann had warned of since the 1930s, served him – and fellow Zionists – as a decisive argument in favor of creating a Jewish state. In his testimony before the Anglo-American Committee in 1946 – yet another international commission of inquiry – Weizmann declared the establishment of a Jewish state to be "the line of least injustice"; to be sure, a measure of political hurt would be done to Palestine's Arabs, but this paled into insignificance beside the hurt done to the Jews and the boon that the state's creation would afford the Arab peoples around Palestine. Besides, the Arabs had just been awarded a cluster of independent states, including in Transjordan, Lebanon, and Syria.[58] Weizmann's argument, in effect, was that any solution to the Palestine problem would cause injustice to one side or the other (or to both); the question was which solution would cause less injustice or assure as little injustice as possible. The wrong done to the Arabs by the state's creation would be far smaller than the damage incurred by the Jews, not only in the recent past, but also that which would be done to them should they fail to receive statehood in Palestine. Put another way, given that the Arab peoples had achieved sovereignty in a large number of states, it was only fair that the Jews receive sovereignty in one (small) state, even if it was established at the expense of Palestine's Arab inhabitants. One member of the 12-man Anglo-American Committee, the British MP R.H.S. Crossman, was struck by the "magnificent mixture of passion and scientific

57 *Trial and Error*, p. 529.

58 "We Warned You, Gentlemen: Testimony of Dr. Chaim Weizmann before the Anglo-American Committee of Inquiry on Palestine," Jerusalem: Jewish Agency for Palestine, 1946, p. 43. I am indebted to B.Z. Kedar for this reference.

detachment" and the "scientific objectivity" of Weizmann's testimony before the committee, a reflection of the mix of passion and scientific thinking that characterized Weizmann throughout his political life.[59]

Weizmann and the Transfer Idea

The idea of removing some Arabs or all the Arabs from Palestine – or from that part of Palestine earmarked for Jewish statehood – in order to facilitate Jewish settlement or to achieve a Jewish majority, had hovered in the Zionist imagination since Theodor Herzl and Israel Zangwill. The idea received a great push in the 1930s, and it was adopted by Zionist leaders, including Weizmann and Ben-Gurion. Weizmann saw the idea as one way of solving "the Arab question." As with others, Weizmann's adoption of the transfer idea was a reaction to the Arabs' intentions and violence toward the Jews and to the deeds of the Nazis in Europe.

Soon after the August 1929 Wailing Wall riots, Weizmann proposed, in talks with Colonial Secretary Lord Passfield (Sidney Webb), to transfer Arabs from Palestine to Transjordan or Iraq. If Britain persisted in opposing the settlement of Jews in Transjordan, perhaps Arabs from Palestine should be allowed or encouraged to settle there. According to Weizmann, Passfield, no friend of Zionism, responded by saying that

> he would have to consider a solution in that direction; but Iraq might present some difficulties ..., and [the Iraqis] were very difficult

59 Richard Crossman, *Palestine Mission: A Personal Record*, New York–London: Harper & Brothers, 1947, pp. 123–124. Walter Laqueur also noted Weizmann's successful mix of scientific argumentation and passion ("irrefutable logic and appeal to emotion") and remarked: "He was probably the most persuasive man I ever met"; idem, *Worlds Ago: A Memoir of My Journeying Years*, New York: Transaction Publishers, 2016, p. 215. When working as a journalist in Palestine in the late 1940s and early 1950s, Laqueur heard, met, and talked a number of times with Weizmann, who was then already in steady decline.

> people. My reply was: "Of course, it isn't easy, but these countries [i.e., Transjordan and Iraq] have to be developed." … Lord Passfield thought this a wide outlook and one to be taken into consideration very seriously.[60]

The idea of Iraq as a potential venue for the transfer of Palestinian Arab peasants may have been planted in Weizmann's mind by the chairman of the Jewish Agency, Colonel Frederick Kisch, who already before the 1929 riots had looked to Iraq in this context. In September 1928, he had informed Weizmann that the director of the Lands Department in the Iraqi British Mandate administration had highlighted the country's "growing need for *fellaheen*" to develop its agriculture. Kisch told Weizmann:

> This is what I have always been hoping and waiting for, having before now expressed the view that with the irrigation of Mesopotamia will be found the solution of the racial problem in Palestine. Propaganda for an Arab emigration [from Palestine] to Iraq cannot be made by us … but for the Iraqi Government to do so would be wonderful, and in the interests of both countries.[61]

In May 1930, Kisch returned to this theme. He wrote Weizmann that the Jewish Agency should press Britain to promote Arab emigration to Iraq.

> It should not be impossible to come to an arrangement with [King] Faisal by which he would take the initiative in offering good openings for [Palestine] Arab immigrants. … There can be no conceivable hardship for Palestinian Arabs – a nomadic and semi-nomadic people – to move to another Arab country where there are better opportunities for an agricultural life.

60 Weizmann (London), "Awaiting the Shaw Report," March 6, 1930, *CWP*, I, no. 116, pp. 591–592.

61 Kisch to Weizmann, September 21, 1928, Kisch Papers, Ben-Gurion Archive.

Kisch pointed to English emigration to Canada as an enlightened precedent.[62]

The outbreak of the Arab Revolt in 1936 and the recommendations of the Peel Commission the following year, for partitioning Palestine into two states and transferring 200,000–300,000 Arabs from the area of the Jewish state-to-be to that of the Arab state-to-be, or out of Palestine altogether, only strengthened Weizmann's support for the transfer idea. He regarded its implementation as of "crucial importance" to "the success of … [the whole] partition scheme," as he wrote to Colonial Secretary Ormsby-Gore.[63] But in his letters, at least to non-Zionists, he endorsed an agreed, voluntary transfer, usually implying, and sometimes stating explicitly, that he opposed a compulsory transfer. (The Peel Commission had proposed a voluntary transfer – but endorsed a compulsory transfer if the Arabs refused to move voluntarily.) In mid-August, Weizmann wrote the president of the League of Nations Permanent Mandates Commission, Pierre Orts: "Of course we do not propose to have recourse to constraint, or to exercise coercion whatsoever: Only those who wish will be transferred, and those [Arabs in the designated Jewish state area] who prefer to stay will stay."[64]

That month, as Weizmann partnered with Ben-Gurion at the emergency Twentieth Zionist Congress in Zurich in leading the pro-partition mainstream, he full-throatedly endorsed the Peel Commission's transfer recommendation.[65] A few days earlier, Weizmann had met

62 Kisch to Weizmann, May 20, 1930, Kisch Papers, Ben-Gurion Archive.

63 Weizmann to Ormsby-Gore, *CWL*, XVIII, no. 159 (July 20, 1937), p. 180.

64 Weizmann to Orts, *ibid.*, no. 168 (August 14, 1937), p. 186.

65 Benny Morris, *Tiqun Ta'ut: Jews and Arabs in the Land of Israel, 1936–1956*, Tel Aviv: Am Oved, 2000, pp. 47–48 (Hebrew). According to Moshe Glickson, also a delegate at the Congress, Weizmann said at the Congress that "in his opinion it was possible to transfer over a 20 year period 100,000 Arabs to the Arab state. … Dr. Weizmann told us about a plan to create a fund for large scale settlement [on behalf of the transferred Arabs]." Menahem Ussishkin, the chairman of the Jewish National Fund, said at the Congress that Weizmann had spoken of the transfer of "300,000 Arabs out of the Jewish state." Weizmann's statements on the matter are not included in the official protocols of the Congress sessions.

Ormsby-Gore. "I said that the whole success of [the partition] scheme depended on whether the Government … [carried] out this [transfer] recommendation," he reported. According to Weizmann, Ormsby-Gore "agreed that once the Galilee was given to the Jews … the position would be very difficult without transfer."[66]

But in September 1937, the Arabs renewed the revolt, and, within weeks, the British government withdrew its support for partition and, with it, the transfer idea. The Zionist leadership continued to discuss the idea, albeit behind closed doors, including in gatherings of the Jewish Agency Executive, the de facto "government" of the Yishuv. In public, the Zionist leaders usually held their peace regarding the transfer idea. But it is clear that Weizmann continued to favor it. For example, in 1938 he supported the transfer of 50,000 Galilee Arabs, of whom he said that some were "utterly impoverished and landless."[67] In April 1939, he proposed the transfer of 10,000 Druse from the Galilee (in other words, most if not all of Palestine's Druse) to southern Syria, so that "all the district along the Acre–Safad road as far as Hanitah and Nahariah would become Jewish in a perfectly peaceful manner."[68] Nothing came of this. Weizmann also endorsed the transfer component within the context of the initiative by Harry St. John Philby, adviser of Ibn Saud, the king of Saudi Arabia, to create a Middle Eastern federation in which a Jewish state in Palestine, denuded of its Arab inhabitants, would figure prominently.[69] Needless to say, nothing came of this initiative either.

Transfer was still exercising Weizmann's mind in 1941 – long after the British had abandoned partition and the idea of transfer – just before

66 Chaim Weizmann, "Summary Note of Interview with Mr. Ormsby-Gore, Colonial Office, Monday, July 19th, 1937, at 10:45 a.m.," CWA, 27-2000.

67 Weizmann to Sami Gunzberg (Istanbul), *CWL*, XVIII, no. 379 (August 2, 1938), pp. 439–440.

68 Weizmann to Solomon Goldman (Chicago), *CWL*, XIX, no. 52 (April 28, 1939), p. 54.

69 See Moshe Sharett, *Political Diary*, IV, ed. Ahuvya Malkin, Tel Aviv: Am Oved, 1974, pp. 374–376 (Hebrew); and Yehoshua Porath, *In Search of Arab Unity 1930–1945*, London: Frank Cass, 1986, pp. 80–106.

the beginning of the Holocaust, though months after the Germans had started herding eastern Europe's Jews into ghettos and killing them. It came up at Weizmann's meeting with Soviet ambassador Maiskii at the end of January, on which both men's reports are available. Weizmann had initiated the meeting in order to propose a barter agreement – oranges from the Yishuv (the European market now being inaccessible because of the war) in exchange for furs from Russia (which perhaps could be stored by the Yishuv until the war's end and then re-exported to the West). After Maiskii politely turned down this proposal, the two discussed other topics – the plight of Europe's Jews, the Palestine problem and the transfer idea. According to Weizmann, Maiskii said that "there would have to be an exchange of population" if the Zionists wished to move masses of European Jews to Palestine. Weizmann responded that if half a million Arabs were transferred out of Palestine, "two million Jews could be put in their place. That, of course, would be the first installment." According to Maiskii's report to Moscow, it was Weizmann who initiated the discussion of transfer, saying that in order to save European Jewry, it would be necessary "to move a million Arabs who are now in Palestine to Iraq, and to settle five million Jews from Poland and other countries on the land where these Arabs were." According to Weizmann, Maiskii objected that moving a "hill-country population to the plains" might be problematic; he also questioned whether five million Jews could be settled in a territory previously inhabited by only one million Arabs. Weizmann – according to his own account – replied that "a beginning might be made with the Arabs from the Jordan Valley" and that "conditions in Transjordan were not so very different from those of the Palestine hill-country." Weizmann then referred, as we saw above, to the Arab propensity to turn gardens into deserts. "Give me the land occupied by a million Arabs, and I will easily settle five times that number of Jews on it," Maiskii recorded Weizmann as saying.[70]

70 For the two reports, see above, note 6.

1948

After hundreds of thousands of Arabs fled or were uprooted from their homes in the course of the 1948 War, Weizmann adopted the official Israeli line – that the refugees would not be allowed to return. Foreign Minister Moshe Shertok (Sharett) wrote him in August that year:

> With regard to the refugees, we are determined to be adamant while the war lasts. Once the return tide starts, it will be impossible to stem it, and it will prove our undoing. As for the future, we are equally determined ... to explore all possibilities of getting rid, once and for all, of the huge Arab minority which originally threatened us. ... What such permanent resettlement of "Israeli" Arabs in the neighbouring territories will mean in terms of making surplus land available in Israel for the settlement of our own people requires no emphasis.[71]

Weizmann apparently needed little convincing. On December 9, 1948, he summoned Yosef Weitz, the director of the Jewish National Fund's Lands Department and an important Zionist executive, to his Rehovot home. In that meeting, he expressed his complete agreement with Weitz's views on the possibility of a return of Palestinian refugees. Weitz, a major proponent of transfer in the 1930s and 1940s, adamantly opposed a refugee return and was busy orchestrating concrete measures to bar one, by destroying villages, cultivating Arab fields, settling Jews in empty villages, and so on. Weitz jotted down in his diary:

> I couldn't believe it. He [Weizmann] was very agreeable. His first question – now you have sufficient land [for your plans to develop the Yishuv]? And this gave me the excuse to lecture him on my views on the Arab question[,] as it had appeared [to me] that he had another [contrary] view[,] conciliatory and compliant. And I was astonished and happy as this was not the case. He accepted my views

71 Shertok to Weizmann (Montreux), July 22/August 22, 1948, in *Documents on the Foreign Policy of Israel*, I: *14 May–30 September 1948*, ed. Yehoshua Freundlich, Jerusalem: Israel State Archives, 1981, p. 369 (Hebrew and English).

on all points. The Arabs cannot and should not return and we should compensate them for their property, in order that they resettle in the Arab states.[72]

But, while opposing a refugee return, Weizmann was highly critical of the Israeli military's excesses against Arab civilians in the course of the fighting. His reservations about Jewish behavior toward the Arabs had surfaced already during the thirties, during the Arab Revolt, when the Irgun Zva'i Leumi (IZL – the National Military Organization – or, as the British called it, the Irgun) began to retaliate with terror attacks against the Arabs. Weizmann condemned the IZL and terrorism in principle. In 1944, when Menachem Begin became the commander of the IZL and launched his "rebellion" against the British, Weizmann called him "a megalomaniac suffering from a Messianic complex. Whether he is a fanatic or a charlatan or both is difficult to say."[73] But Weizmann understood that the problem of Jewish behavior toward the Arabs was not restricted to the IZL and that the organization's behavior was symptomatic of a more general change in the Yishuv's moral compass.

During a visit to Palestine in 1945, Weizmann painted "a dark picture of [the] psychology of [the] rising generation in the Yishuv [and] said

72 Diary of Yosef Weitz, Central Zionist Archive, A246-14, p. 2544, entry for December 9, 1948. Two decades later, in the published version of his diary, Weitz rephrased the entry thus: "I lectured [Weizmann] on the Arab question. … I was fearful lest he view [my views] as extreme, but how great was my joy when he said that he agreed with me on all points. The Arabs could not and should not return. We shall compensate them for their property, so that they can resettle in the Arab states." Weitz, *My Diary and Letters to My Children*, Ramat Gan: Masada, 1965, III, p. 363 (Hebrew). Four months later, according to Weitz, Weizmann again asked him what was happening regarding the Arab question. Weitz recorded: "He has not yet read our memorandum on solving the refugee [issue in the Arab states]. He also took an interest in the new settlement [of Jews on abandoned Arab lands]. I told him about the settlement of new immigrants on the land. He listened keenly, but without reacting. We parted." *Ibid.*, IV, pp. 17–16, entry for March 9, 1949. In this case the original handwritten entry in the diary kept in the CZA is not substantively different; see A246-14, pp. 2594–2595.

73 Rose, *Chaim Weizmann* (above, note 3), p. 399.

[that] Ben-Gurion is much to blame."[74] The Zionist slide into militancy and eventual "militarism," in Weizmann's phrase, may have found its wellsprings and justification in Arab terrorism in Palestine and Nazi mass murder in Europe. But it mortified Zionism's elder statesman, who was quickly losing his grip on the movement. In his memoirs, he wrote of this period:

> Here and there a relaxation of the old, traditional Zionist purity of ethics, a touch of militarization …; here and there, something worse – the tragic, futile un-Jewish resort to terrorism, a perversion of the purely defensive function of *Haganah*; and worst of all, in certain circles, a readiness to compound [*sic*] with the evil, to play politics with it, to condemn and not to condemn it, to treat it not as the thing it was, namely, an unmitigated curse to the National Home, but as a phenomenon which might have its advantages.[75]

In summer 1946, against the backdrop of the IZL demolition of the south wing of the King David Hotel in Jerusalem, in which some 90 Britons, Arabs, and Jews died, Weizmann wrote to Shertok: "In the eyes of all men of goodwill, our movement is sinking to the level of gangsterdom."[76]

By 1948, according to Weizmann, the gradual shift in the Yishuv's outlook from the defensive to the offensive that had occurred during the Arab Revolt and the Second World War[77] had resulted in "the moloch of militarism" laying hold of the Zionist movement itself. Zionism – "everything and everybody" – was in its grip, he wrote his protégé Meyer Weisgal from Switzerland, where he was recuperating from illness (he arrived in Israel only in late September). Indeed, he wrote, he was

74 *Ibid.*

75 *Trial and Error*, p. 539.

76 Rose, *Chaim Weizmann* (above, note 3), p. 411.

77 See Anita Shapira, *Land and Power: The Zionist Resort to Force, 1881–1948*, New York: Oxford University Press, 1992.

reluctant to return to Israel to assume the mantle of the presidency, a job, he wrote, that had been "foisted upon me":

> I am not in a position to cover with my name all that is going on at present in Palestine and to accept everything which the government does without being able in any way either to influence it or to prevent it. ... Meanwhile, facts are being created which one would have to face, and I am not prepared to do so.

What Weizmann was referring to is not clear, though he may have heard rumors or received information about the IDF massacre of townspeople in Lydda and the expulsion of its inhabitants, along with those of neighboring Ramleh, on July 12–13. Weizmann complained that Israel's leaders were keeping him "entirely in the dark" about what was happening, and he even threatened to resign.[78]

Among the "facts" that deeply disturbed Weizmann during his sojourn outside Palestine in 1948 were the wartime changes at the Sieff – later Weizmann – Institute. He gave vent to his criticism in the same long letter to Weisgal. He complained that he wasn't being given information about "what is really going on there," except that, generally, the institute had switched to a war footing; that no one had consulted him about this; and that "the whole of the scientific work" had been "replaced by something which is not science but making of explosives." In this context, too, he threatened resignation: "I consider myself as the head of a scientific centre, but if this centre changes its character without even as much as 'by your leave,' I find that this relieves me of my duties." He even defined what was happening to the institute as "desecrat[ion]."[79]

The Institute indeed underwent a general transformation during the

78 Weizmann to Meyer Weisgal (New York), *CWL*, XXII, no. 238 (July 30, 1948), p. 192.

79 *Ibid.* See also William B. Jensen, Henry Fenichel and Milton Orchin, *Scientist in the Service of Israel. The Life and Times of Ernst David Bergmann (1903–1975)*, Jerusalem: Magnes Press, 2011, pp. 156 and 168.

war, so that already in May 1948 almost all the scientists and laboratories, working in shifts around the clock, were engaged in producing materials and products geared to the war effort. The Institute also expanded in terms of buildings and laboratories.[80] The pre-war staff of about 100 scientists and technicians was joined by dozens more from the Hebrew University of Jerusalem and the Haifa Technion, and another hundred from the IDF.[81] The Institute had become the country's leading scientific institution. Weisgal described the transformation: "The infant in Rehovot has grown into manhood, with a devouring appetite …"[82] By the end of 1948, some 350 scientists and technicians were employed on war projects in Rehovot and its environs, Weisgal told the Institute's "American Committee."[83]

The scientists and technicians worked on research and production of a wide variety of war-related materials. The most revealing survey of this work (that is open to researchers) was written by Ernst David Bergmann, the Institute's scientific director at the time. In August 1948, he wrote to Weisgal (presumably expecting the information to reach Weizmann) about a few of the projects. He reported on the production of aviation fuel and distilled water, remote control electronic equipment, sniperscopes, fluorescent paints – used to make certain equipment, such as artillery markers, visible at night – explosives, colored smoke, propellants for mortar bombs and rockets, "2.5 kilogram" rockets, and materials to neutralize poison gas, and on testing metals for weapons production. Bergmann hinted at several other items in referring to a project managed by Dr. Israel Dostrovsky and Dr. Herbert Bernstein, work that he defined as crucial to the war effort, which involved

80 Weisgal (Tel Aviv) to Weizmann, CWA 2824-30 (April 14, 1948). Weisgal defined the expansion of the Institute as "breathtaking."

81 Ernst David Bergmann, "Report on the Scientific Activities of the Weizmann Institute of Science during the Period January 1948–July 1950," WIA, 9.49 (July 1950).

82 Weisgal to Weizmann (Rehovot), CWA, 2876-24 (December 10, 1948).

83 Weisgal at the Annual Meeting of the American Committee for the Weizmann Institute of Science," Plaza Hotel, New York, December 8, 1948, WIA, protocols of the Board of Governors, 1948, 44.9.bet.

experiments that left the scientists with "painfully inflamed eyes." He appears to link this project – without explanation – to the capture of police forts.[84] He may have been referring to the development of a tear gas, on which Aharon Katchalsky (Katzir), who worked at the Institute, reported to Ben-Gurion on February 2, 1948: "'Shimshon' … an experiment was carried out on animals. The researchers were dressed in suits and gas masks. … No animal died; the dazzling/blinding effect [*sinvur*] lasted 24 hours."[85]

Bergmann may have been referring to this line of research in writing of the work of a group of scientists headed by Ephraim Katchalsky, Aharon's brother, at the laboratories of the Hebrew University's nearby agricultural research station, which were co-opted by the Institute during the war. That group, he wrote,

> is concerned mainly with the study of the factors influencing the dispersal of active substances in the air. The techniques we are studying are, to a large extent, taken from the experience gained with insecticides – but the substances are far from being insecticides only.[86]

As far as is known, the Haganah and the IDF did not use poison gas during the war. But in summer 1948 the Yishuv feared that it might be used by the Arab armies, including Jordan's Arab Legion – a fear that

84 Bergmann to Weisgal, WIA, Meyer Weisgal, Correspondence, 1948, 8 (August 11, 1948). In another letter, Bergmann wrote that the Nabi Yusha police fort in the Galilee Panhandle was captured "solely by the application of one of our little devices"; Bergmann to Weizmann, CWA, 2832-6 (May 18, 1948), quoted in part in Jensen, Fenichel and Orchin, *Scientist* (above, note 79), p. 139. It is possible that the reference was to an explosive device used to blow a hole through the fort's perimeter wall. See also Bergmann to Weizmann, CWA, 2847-4 and 2851-4 (July 6 and July 26, 1948).

85 David Ben-Gurion, *The War Diary: The War of Independence, 1948–1949*, eds. Gershon Rivlin and Elhanan Orren, Tel Aviv: Society for the Dissemination of Ben-Gurion's Legacy – Ministry of Defense, 1983, Part 1, p. 201 (Hebrew).

86 Bergman to Weisgal, August 11, 1948 (above, note 84).

triggered the production of neutralizing elements or antidotes.[87] But in the end, the Arab armies also made no use of poison gas.

One last thing: In summer 1948 a number of Institute scientists began work – at least part-time and apparently only on a theoretical level – on military aspects of nuclear research. On July 26, Bergmann wrote Weizmann that he no doubt would be "pleased to hear that we are beginning to work on heavy water in connection with one of the projects which you have discussed so frequently with me." Bergmann spoke of thorium deposits in Egypt,

> and very probably also in this country; we do not think it impossible to imitate the work done with uranium. Prof. [Giulio Yoel] Racah of the Hebrew University has been appointed the head of this research group, and Dr. Dostrovsky of our Institute will be one of the main contributors to this project. This is the kind of work which is of military nature but still might fit into the frame of our Institute.[88]

It is unlikely that the Institute's rapid shift from peaceful research to a war footing and its massive mobilization for the war effort in the first months of 1948 overly perturbed Weizmann. He was no pacifist: In World War I, he famously subordinated his research to the British war effort and orchestrated the production of materials for munitions; in 1941, he objected that the Institute had thus far done nothing on behalf of the Allied war effort;[89] and he regarded the war against the Arabs in 1947–1948 as a just war. Perhaps Weizmann was ired by the wholesale shift from peaceful research to applied production for the war effort, which meant abandonment or suspension of certain avenues of research that he had promoted. Perhaps he had heard rumors or facts about the

87 Bergman to Weizmann, July 6, 1948 (above, note 84).

88 Bergmann to Weizmann, July 26, 1948 (above, note 84). See also Ben-Gurion, *War Diary* (above, note 85), Part 3, p. 827. Racah and Dostrovsky were later among the builders of the Dimona nuclear reactor and served on the Israel Atomic Energy Commission.

89 Jensen, Fenichel and Orchin, *Scientist* (above, note 79), p. 111.

production of certain chemical or biological agents of which he did not approve. Or perhaps he was afraid that thenceforward the Institute would serve mainly as a military research center – something that Bergmann had hinted at in one of his letters.[90]

But without doubt Weizmann was angered by his dissociation from the Institute and its work and the divorce that developed between himself and his favorite scientific protégé, Bergmann – his right-hand man before the war – in the war's first months. He repeatedly complained that Bergmann failed to keep him updated about developments in the Institute or to consult him;[91] in effect, Bergmann appended himself and the Institute to Ben-Gurion and the Haganah Science Department, later the IDF's Science Corps. During 1947–1948, Ben-Gurion mobilized the Hebrew University and the Haifa Technion, the country's two institutes

90 Bergmann had written Weizmann: "What I would like to do is to concentrate on only that pure research in the Institute which will be of vital importance in any future [military?] emergencies"; Bergmann to Weizmann, July 26, 1948 (above, note 84). Bergmann was convinced that Israel had to develop military industries and autarkic production capabilities, and in another letter to Weizmann he listed planning for the independent production of weapons and military materials as a chief priority, ahead of national planning for industry and agriculture in general. He also recommended that the personnel of the IDF's Science Corps serve as the nucleus of the body that in the future would chart out the country's scientific development; Bergmann to Weizmann, CWA, 2841-16 (June 6, 1948). See also Jensen, Fenichel and Orchin, *Scientist* (above, note 79), pp. 145–146 and 160–162.

91 See, e.g., Weizmann (Vevey, Switzerland) to Weisgal (New York), *CWL*, XXIII, no. 242 (August 3, 1948), pp. 197 and 199. It is not clear from the letters in the CWA and the WIA why Weizmann complained of being kept out of the loop by Bergmann. During spring–summer 1948 Bergmann sent him a great many letters, one per week on average, many of them long and highly detailed. But in one of the letters Bergmann wrote that the Institute's persistent past focus on theoretical and civilian research may have been a protracted mistake: "I often wonder whether we have done the right thing in doing in the Institute the sort of research we have carried out in these fourteen years, instead of preparing the country chemically and mechanically for the most refined of modern warfare methods"; Bergmann to Weizmann, CWA, 2839-17 (May 31, 1948). Weizmann likely did not take kindly to this severe, albeit implicit, criticism. [See Shaul Katz's article in this volume, which discusses the rupture at length. – Ed.]

of higher education, for the Yishuv's defense needs, and Weizmann apparently viewed the shift by Bergmann and the Institute at Rehovot to war production as a shift to the "Ben-Gurion camp," a type of betrayal.[92]

Weizmann no doubt also viewed what had happened at the Institute within the context of his long struggle with Ben-Gurion over the leadership of the Zionist movement.[93] Weisgal tried to calm his anger on this count in his letter of August 13:

> I don't think it is the intention of anyone to subvert the original intentions of the Institute. When normal conditions return, we too will get back to normal activities in Rehovot. I would only ask you to please not be too hard on any of the people either in Tel Aviv or Rehovot."

The allusion is probably to Ben-Gurion and Bergmann.[94] But Weizmann stuck to his guns:

> Nothing which has happened in the last few years of my life has hurt me so much. … For the time being, I am afraid I do not see any solution to the problem except, perhaps, to withdraw quietly from the Institute.[95]

Weisgal's effort at calming the waters had failed.

Another issue that exercised Weizmann in 1948 relates to the IDF's behavior toward Palestine's Arabs. Powerful evidence in this connection is to be found in the unpublished diary of Yosef Nahmani, a Second Aliya figure who in 1948 was a member of the municipal council of Tiberias and the JNF branch manager in eastern Galilee. He was

92 Jensen, Fenichel and Orchin make this point in *Scientist* (above, note 79), p. 176.

93 I would like to thank Dr. Uri Cohen, who is writing a book on the history of the Sieff Institute, for discussing this interpretation, or something like it, with me in a telephone conversation on March 30, 2014.

94 Weisgal to Weizmann CWA, 2854-15 (August 13, 1948).

95 Weizman (Burgenstock) to Weisgal (Rehovot), *CWL*, XXIII, no. 325 (August 18, 1949), p. 279.

something of an expert on Arab affairs. Officers and officials kept Nahmani abreast of what was happening. He was summoned to meet Weizmann on November 13, 1948, at Tiberias's Galei Kinneret Hotel, where the President was vacationing. This was a fortnight after the end of Operation Hiram, in which the IDF conquered central–northern Galilee, from the ʿArabe–Sakhnin–ʿEilabun line up to the Lebanese border.

According to Nahmani, Weizmann opened the conversation: "We achieved our state in our lifetime, a great miracle happened here." But, he added, "I heard that the Arabs who stayed are being treated badly, I have heard of acts of robbery, rape, and oppression … and this will rebound on us. We will be judged by our behavior and moral values." Nahmani apparently told Weizmann of the series of massacres perpetrated by IDF soldiers during Operation Hiram and in its immediate wake.

> "Where did our sons learn such cruelty? How did this happen?" Weizmann asked, and added: "Aren't you consulted about how to behave toward the Arabs?"
> Nahmani answered: "They don't ask me. And do they ask you, who are the president?"
> "No," he [Weizmann] answered me in a despairing, despondent voice. "No … they don't ask me."
> Nahmani concluded: "Weizmann was very depressed, looked sad, and his appearance reminded me of when I met him in Jaffa and Jerusalem during the first [anti-Jewish] riots in the country [in 1920], when he was on his way to meet the high commissioner."[96]

96 Diary of Yosef Nahmani, Hashomer Archive, Kefar Gil'adi, entry for November 13, 1948. There is a slight chronological problem here. The first anti-Jewish riots occurred in Palestine in April 1920, a few months before the first British high commissioner, Herbert Samuel, arrived in the country. Hence, either the early meeting Nahmani mentions occurred not during the April riots but months later, or perhaps during the second bout of rioting, in May 1921, or else, if it indeed took place during the April 1920 riots, then Weizmann was not on his way to see the high commissioner.

The two men met again the following day, and according to Nahmani the conversation developed as follows:

> Weizmann: "I couldn't calm down all night because of our boys' acts against the Arabs. What can be done to prevent condemnable acts and depredations?"
> Nahmani: "My friend the president … only the government has the power to restrain the instincts vented by some of our boys."
> Weizmann: "I am afraid … that I have no influence. They [Ben-Gurion and company?] make sure that I have no influence."
> Nahmani: "His words were very sad. After a 29-minute conversation I left him, after promising that I would try to visit him in Rehovot from time to time."[97]

But Weizmann's reactions to what was happening in 1948 were multi-layered. When speaking to fellow Zionists, he was often critical of Israel's behavior. But when corresponding with non-Jews, he criticized the Arabs' behavior. He condemned the Arabs' rejection of the United Nations partition plan and their launching of hostilities against the Yishuv, criticized Britain for its support of the Arabs while knowing that the Arabs sought to slaughter the Jews,[98] and mocked the Arab armies' performance and the Palestinian Arabs' flight when the Jews neared their homes.[99] A certain belligerence apparently characterized his views when speaking in intimate circles. According to his nephew Ezer Weizman, an IDF pilot (and later also president of Israel), Weizmann asked him, after the IDF's penctration of Sinai at the end of December 1948: "Why aren't we pushing on to Cairo?"[100]

Upon returning to Israel in late September 1948, Weizmann went into a propagandistic, state-serving mode. To Walter Elliot, a British MP

97 *Ibid.*, entry for November 14, 1948.

98 Weizmann (Geneva) to Leonard Stein (London), *CWL*, XXIII, no. 252 (September 17, 1948), p. 206.

99 Weizmann to Winston Churchill, *ibid.*, no. 246 (August 6, 1948), p. 202.

100 Rose, *Chaim Weizmann* (above, note 3), 448.

and former cabinet minister, he wrote: "The Arabs who have chosen to stay in the country are contented"; by contrast, "those who left" were "not welcome" in the neighboring Arab countries. "I do not know who advised the Arabs to run away," he continued, implying that British officials had tendered such advice; "it was not merely fear that drove them out; they were not subjected to either threats or pressure from our side." This was at best a quarter-truth, and Weizmann could not but have known that. And surely he could not wholly have believed his following assertions, in which he implied that it was the Arabs – specifically, Haj Amin al-Husseini – who had ordered the Arabs to leave their homes, thus creating the Palestinian refugee problem: "I understand that the Mufti was anxious to collect the Arabs in one place and form them into an army which would attack and destroy Israel. ... We did nothing to cause this mass flight."[101]

In two lengthy letters, to Lady Violet Bonham-Carter, daughter of Herbert Asquith, Britain's prime minister during the first years of World War I, and to U.S. President Harry Truman, Weizmann defended the policies of the Yishuv and the new state vis-à-vis the British and the Arabs. Lady Bonham-Carter had written to him implying that the Jews had been guilty of anti-British terrorism and murder before 1948 and, subsequently, of inhumanity toward the Palestinians. She condemned Israel for creating the Palestinian refugee problem.

Weizmann responded:

> The Jews are not a people given to violence. ... Our colonization in Palestine was an outstanding achievement of non-violence. We maintained the peace even when our fields and settlements were set on fire, and peaceful men, women and children were murdered.

But Britain's appeasement of the Arabs following the three years of Arab rebellion in the late 1930s had led some Jews to believe that "violence paid with the British Government." Then Britain shut Palestine's doors to escapees and survivors of the Nazi Holocaust.

101 Weizmann (Rehovot) to Elliot (London), CWA, 2875-12 (December 1, 1948).

> You are surprised and disgusted that we have terrorists … [But] any nation which would have to go through the cruel ordeal which the Jews have passed under Hitler would have produced such growth in terrorism and even worse.

He implied that Bonham-Carter's "self-righteous indignation" was in some measure caused by a sense of guilt.

As for the "800,000" Palestinian refugees "ravaged by starvation, disease and death," as she had put it, Weizmann wrote: "The actual number is about half that figure." In any event, he asked, "whose responsibility is it?"

> Did we declare war on the Arabs and drive them out of the country? Was it not rather the Arab states that invaded our country on the 15th of May and even before …? The poor Palestine Arabs were misled by promises that the Jews will be exterminated in a very short while and then they would be brought back to what they had left, by the victorious armies of Arabistan led by a good few British officers. It was not part of our plan of campaign to dispossess the Arabs. When the flight started – long before the British Administration left Palestine – we tried hard to persuade them to stay.[102]

Two months later, Weizmann wrote to Truman, elaborating that:

> these peoples [*sic*] are not refugees in the sense in which that term has been sanctified by the martyrdom of millions in Europe – they are part of an aggressor group which failed and which makes no secret of its intention to resume aggression. [The refugees] left the country last year at the bidding of their leaders and military commanders and as part of the Arab strategic plan.

102 Weizmann to Bonham-Carter, *CWL*, XXIII, no. 308 (March 13, 1949), pp. 262–265. Needless to say, some of what Weizmann wrote about the creation of the Palestinian refugee problem was accurate and some of it grossly inaccurate.

In any event, added Weizmann, Israel had been "steadily re-admitting Arab refugees during the last few months" and would re-admit more as part of a peace settlement, but only to the extent that it did not impair Israel's security or economic well-being. *En large*, the refugee problem had to be resolved through resettlement in the Arab states, Weizmann believed.[103]

Weizmann died in 1952. It is unclear whether he believed that peace would ever reign between Israel and the Arab states. But he believed that the State of Israel should behave fairly and with an egalitarian approach toward its Arab citizens. Referring to Leviticus 24:22, he wrote in his autobiography: "There must not be one law for the Jew and another for the Arabs … [the Arabs] must be treated exactly like the Jews."[104] As to Israel's position in the world, Weizmann understood that the West's dependence on Arab oil harmed Israel's position and ate away at its security, and he devoted some space to this in his autobiography. He hoped and believed that Israel would become "a center of the new scientific development which would get the world past the conflict arising from the monopolistic position of oil."

Weizmann believed that the production of substitutes for important oil products by processes of fermentation of field crops, an area to which he devoted a great part of his scientific life, might help lead to the liberation of the West from dependence on Arab oil. He envisioned the Weizmann Institute as being at the heart of scientific research in this field in Israel.[105]

What emerges from this study of Weizmann's attitudes regarding the Arabs and Zionist policy toward them is a complex and changing picture: From his initial Old World attitude toward the "natives" and non-recognition of the Palestine Arabs as a distinct people, he moved toward a grudging acceptance of the rise of a separate Palestinian Arab

103 Weizmann to Truman, *ibid.*, Appendix 1, pp. 314–316 (June 24, 1949).

104 *Trial and Error*, pp. 566 and 569–570.

105 [See also Section IV in the chapter "Unpublished Passages" in this volume. – Ed.]

nationalism, side-by-side with a condemnation of that nationalism as aggressive and fascistic. In the end, Weizmann resigned himself to the idea of partition and a two-state solution.

From the start, Weizmann recognized the indigenous population's right to permanent habitation and to security of person and property. But in the 1930s and the early 1940s, in the context of the Arab assault on the Jews in Palestine and the persecution of the Jews in central and eastern Europe, which necessitated the establishment of a safe haven in the Land of Israel, he put aside his misgivings regarding the eviction of populations and supported the transfer of the Arabs from the area earmarked for Jewish statehood. In the course of the 1948 war, Weizmann expressed no objections to uprooting Arabs from the areas conquered by the Haganah and the IDF and supported barring the return of the refugees. But as a liberal and enlightened European, he was deeply critical of the Jews' behavior toward Arab civilians when it culminated in acts of massacre and depredation.

Nevertheless, as the most prominent spokesman of Zionism, Weizmann did not express his reservations and condemnations in public and persisted in loudly defending Israel's actions in the critical months of its emergence. One might say that, from the 1930s on, Weizmann's attitude toward the conflict was embodied in his saying that the Zionists must adopt "the line of least injustice" and reach a settlement with the Arabs based on this principle.

Translated into English by the author

Raphael Mechoulam

Weizmann: Applied Science and Patents

From the start of his academic career, Chaim Weizmann engaged in both basic and applied chemical research. During his studies and while serving as a *Privatdozent* in Geneva, he sold two patents that had grown out of his work on dyes, to the German firm I.G. Farben and to a French company. Of the former, he wrote proudly in his autobiography: "This provided me at once with a regular income of six hundred marks a month. That was a tremendous experience for me. I had become independent! What was more, I had achieved independence by my own efforts, and in my own field as a chemist."[1] Today it is difficult to evaluate what significance these patents had for the European dye industry, which was then developing rapidly. The huge number of patents for dyes registered in just a few decades around the turn of the century makes it nearly impossible to judge what contribution each individual discovery made to the field.

Weizmann's most important scientific discoveries were in biotechnology, a field – named only later – in which he was an early pioneer. As he wrote,

> My interest in biological chemistry and in bacteriology as a special branch of organic chemistry began some years after I had settled in Manchester. Facilities for this work were lacking at the university, where biochemistry did not form part of the curriculum at that time, while the study of bacteriology was confined to the medical school. I began to pay frequent visits to the Pasteur Institute in Paris, where I

1 *Trial and Error*, p. 77.

> worked in the bacteriological and microbiological departments. For a time I devoted most of my holidays, Christmas, Easter and summer, to these interests.[2]

In fact, he was quite modest about the huge contribution he made to the development of an entirely new branch of research: Microorganisms had not yet been used to synthesize organic compounds in any organic chemistry laboratories at the time.

Weizmann first attempted to obtain isobutyl alcohol through the fermentation of starch or corn. He conjectured that he could convert this alcohol into isoprene and then, via polymerization, produce an analogue of natural rubber. After testing a large number of bacteria, he discovered a new one, later named *Clostridium acetobutylicum*, which produced excellent results. He wrote:

> In the course of this investigation I found a bacterium which produced considerable amounts of a liquid smelling very much like isoamyl alcohol. But when I distilled it, it turned out to be a mixture of acetone and butyl alcohol in very pure form. Professor Perkin advised me to pour the stuff down the sink, but I retorted that no pure chemical is useless or ought to be thrown away."[3]

Weizmann continued to study these reactions and obtained a high yield of acetone and butyl alcohol. He registered patents for both processes, both of which became and remain highly important for manufacturing. Indeed, just recently, a century after Weizmann discovered this fermentation process, a Brazilian biotechnology company, Gran Bio, signed a contract with an important European company, Solway, to construct a facility for producing butyl alcohol from sugarcane straw, a waste product![4]

At the beginning of World War I, Weizmann offered these fermentation

2 *Ibid.*, p. 171.

3 *Ibid.*, p. 172.

4 *Chemical & Engineering News*, Aug. 19, 2013.

processes to the British War Office, but his offer was taken up only two years later, when First Lord of the Admiralty Winston Churchill asked Weizmann to produce huge amounts of acetone, a vital ingredient in the composition of explosives. Weizmann thus became, apparently, the first scientist to take an innovative microbial fermentation reaction out of the laboratory and implement it on an industrial scale. At a later stage of the war, the acetone production project was transferred to Canada and the United States, as well as to India, where rice was used as a raw material. Understandably, Weizmann's patents relating to this were not made public during the war, but when the war was over they were released and developed by Chemical Solvents, an important American chemical firm.

Weizmann did not return to Manchester after the war, and for a period of nearly twenty years, from 1915 to 1934, he published only a single research article, based on his previous studies of butyl alcohol. Most of his energies during this period were in the service of Zionism. He nevertheless maintained a small private laboratory, where he focused on manufacturing development initiatives that were important at the time to industrial chemistry and remain important today.

One of these fields was the production of titanium-based dyes, in which Weizmann registered ten patents from the early 1920s through the early 1930s. Titanium oxide and several titanium derivatives remain important in the dye industry, but it is not clear whether the processes that Weizmann developed were actually used in the manufacture of dyes.

Another branch of research that interested Weizmann was the transformation of raw materials produced by the refinement of crude oil into products such as benzene, toluene, xylene, and naphthalene, as well as the production of volatiles such as butylene and isobutylene. His principal objective was to produce a stock of toluene, the basis for the production of TNT. This process turned out to be highly valuable. Weizmann registered several patents in this area, which were later purchased by a private British company that owned refineries in Manchester.

During World War II, Weizmann resolved, given his success in the biotechnological production of acetone and butyl alcohol, to seek a biological fermentation method of producing high-octane gasoline, especially in demand as airplane fuel. Butylene was one of the chemicals used in the production of high-octane fuel. Butyl alcohol, which Weizmann had produced in his fermentation processes, could easily be used to produce butylene. But Weizmann encountered strong opposition from the oil companies, and his initiative made no progress during the war. "To initiate a process which had not the approval of the oil companies," Weizmann protested, "was almost too much of a task for any human being."[5]

Weizmann's ideas for producing synthetic rubber were taken up many years later. He discovered that butadiene, which could be produced from butyl alcohol, could be purified by forming a crystalline addition product with liquid sulfur dioxide, which could then be polymerized by using small amounts of metallic sodium as a catalyst. During World War II this process was used to make a synthetic rubber called Buna.[6] While the method was not implemented in the United States, Weizmann's knowledge and experience were greatly valued there. Henry Wallace, who served as vice president during the war, wrote: "The world will never know what a significant contribution Weizmann made towards the success of the synthetic rubber program at a time when it was badly bogged down and going too slowly."[7]

Another subject in which Weizmann took great interest, from the late 1930s until the end of his life, was the production of inexpensive, nourishing, and easily digestible food. His idea was to upgrade materials used to feed cattle so that they could serve as human food. Among these were soybeans, nuts, and nut cakes left over after the extraction of oil. He registered relevant patents. According to Weizmann, these products were tested in many hospitals, where they were fed to patients and found healthful. The production costs were low, making them affordable to even the poorest populations. Produced entirely from plants,

5 *Trial and Error*, p. 527.

6 The name is a contraction of the butadiene-natrium, the latter being the German word for sodium.

7 *Trial and Error*, p. 528.

the products were pure and nourishing and tasted much like meat, even though they contained none. Weizmann thus proposed that they could be of great value for countries in which meat was expensive or forbidden for religious reasons.[8]

Chaim Weizmann was a scientist of original and even revolutionary ideas who also knew how to implement them in manufacturing. He made fundamental contributions that lay at the foundation of many industries, especially in the field of biotechnology.

8 [See Weizmann's lengthy exposition of these possibilities in Section IV in the chapter "Unpublished Passages" in this volume. – Ed.]

Raphael Lamed and Edward A. Bayer

Discovery of the Cellulosome
In the Footsteps of Weizmann's Vision for Biofuels

Chaim Weizmann was a founding father of biotechnology, the first scientist to transfer fermentation processes from the laboratory to industry. He invented the first process to employ bacteria on an industrial scale, rather than chemical reagents or petroleum resources, to produce chemical compounds and solvents from starch. Weizmann held more than 120 patents growing out of his work as a chemist, an impressive number then and today.

His best-known scientific discovery – of the bacterium now named after him, *Clostridium acetobutylicum* (the "Weizmann organism") – was published a century ago. The organism's economic value derives from its ability, by fermentation under anaerobic conditions, to produce large quantities of three biological fuels: acetone, ethanol, and butanol. During World War I, the British military used it to produce acetone, a vital ingredient in explosives, of which there was a severe shortage. (Prior to the war, the British imported acetone from Germany – the enemy!) The process also offered the revolutionary potential of manufacturing synthetic rubber. The Weizmann bacterium (which Weizmann's mentor at the University of Manchester, William Henry Perkin, Jr., advised him to "pour down the sink"[1]) thus became legendary during the first decades of the twentieth century.

1 *Trial and Error*, p. 172.

WEIZMANN AND BIOLOGICAL FUEL

Weizmann realized that his organism could be exploited to transform plant products into fuels. But, despite his work, that vision was not realized until recently. He was the first scientist to propose to base the fuel and organic chemical industries on the fermentation of plant biomass rich in starch, cellulose, and other sugars. Weizmann and others abandoned further development of this revolutionary concept at the time because of the availability and low price of natural fuels such as oil and coal, and because of the vehement opposition of the oil industry, which feared competition. But the energy crisis of the 1970s led to renewed interest in Weizmann's ideas.

Today, in particular, people are aware of the rapid depletion of the world's natural fuel reserves and the urgent and inescapable need to develop alternative sources of energy. Furthermore, growing awareness of the detrimental climactic effects of fuel consumption have prompted the search for sources of energy that do not contribute to global warming, in that they either form less carbon dioxide, like hydrogen gas, or recycle the carbon dioxide formed by burning fossil fuels, thus maintaining or reducing atmospheric levels. This has led energy researchers, economic and environmental policy makers, and the larger public to focus on biofuels.

In recent years, with the prices of oil and gas on the rise and the shrinking of global reserves, Weizmann's idea of using biological fermentation to produce fuel from renewable energy resources has now been put into practice in the United States and Brazil, where ethyl alcohol is being produced from starch derived from corn and sugar cane. The ethanol is used as fuel in internal combustion engines in vehicles and for other purposes.

Our research began during the energy crisis of the 1970s. Our scientific goal, then and today, is to develop a process that can turn renewable plant material into liquid biofuel, such as ethanol and butanol. We thus see our research as an organic outgrowth both of Weizmann's scientific vision and, as we will show below, his actual research.

The Jungles of Burma and the Study of Enzymes that Break Down Cellulose

The most common complex carbohydrate is cellulose, an essential component of the rigid cell walls of plants. About half of the refuse produced by human society, piled up in mountainous landfills, consists of cellulose of plant origin. In industry, cellulose is used principally to manufacture paper, but this process produces billions of tons of cellulose waste, which is buried and constitutes a severe environmental blight.

The use of this refuse to produce biofuels would solve an environmental problem while also helping to fill the growing need for alternative energy sources. Furthermore, unlike our shrinking oil and gas reserves, cellulose can be produced from plants, which can be harvested and then grown again the following year.

Cellulose, like starch, is a complex sugar made entirely of glucose, but it is much more intractable than starch to enzymatic hydrolysis. Breaking down cellulose into simple soluble sugars is the first stage in the process by which many organisms convert their food into energy. While the human digestive system is unable to produce energy from cellulose, herbivores, notably ruminants – animals such as cattle, goats, and sheep – can do so. They owe this ability to symbiotic bacteria that live in their digestive systems and are able to break down cellulose into the simple carbohydrate glucose, which can be used as a source of energy.

What mechanisms, processes, and enzymes enable these bacteria to break down cellulose? And can they be used in the industrial production of biofuels? These were our central research questions.

The study of cellulose-digesting enzymes began in the 1940s, during World War II, in the jungles of Burma (today's Myanmar). The tents used by American soldiers stationed in these jungles, made of fabric manufactured from cellulose fiber, were completely eaten away by the fungus *Trichoderma reesei*. Scientists discovered that this fungus digested the fabric by secreting a range of free enzymes that adhere to the fiber and break it down. Each enzyme of this type has a component

Figure 1: Scanning electron micrograph of *Clostridium thermocellum* bacterial cells attached to the cellulose on which they grow.

responsible for attaching it to cellulose and a catalytic component responsible for breaking down the cellulose that comes into contact with it. The fungus then lives off the resulting simple sugars.

THE DISCOVERY OF THE CELLULOSOME

About seventy years after Weizmann discovered his bacterium, we met and decided to collaborate in our research.[2] In the first stage, we decided to characterize the component that enables a cellulose-digesting bacterium to adhere to the material on which it grows.[3] As a research model, we chose *Clostridium thermocellum*, a relative of the Weizmann bacterium. It is a thermophilic (heat-loving) anaerobic bacterium known for

2 Edward A. Bayer and Raphael Lamed, "The Cellulosome Saga: Early History," in *Cellulosome*, ed. Vladimir Uversky & Irina A. Kataeva, New York: Nova Science Publishers Inc., 2006, pp. 11–46.

3 Edward A. Bayer, Rina Kenig, and Raphael Lamed, "Adherence of *Clostridium thermocellum* to Cellulose," *Journal of Bacteriology*, 156 (1983), pp. 818–827.

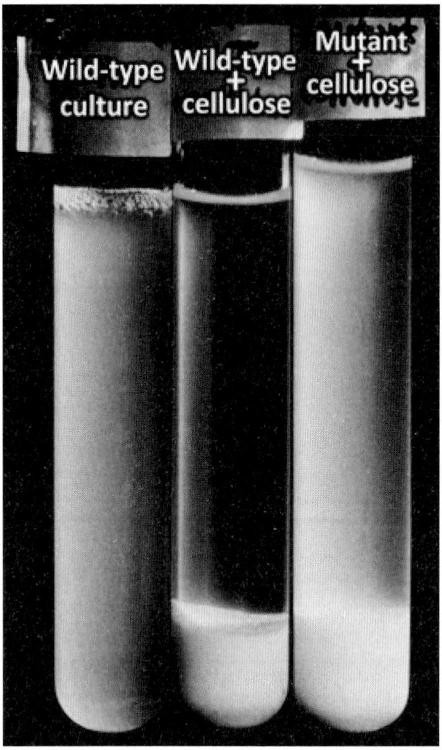

Figure 2: When wild-type bacteria (left) come into contact with cellulose, they adhere to it and settle to the bottom of the test tube (center). The mutant bacteria remain free, suspended in the liquid (right).

its ability to dismantle cellulose by breaking it down into simple sugars, which it then turns into ethanol.

How could the adhesion mechanism be studied? We began by isolating a mutant strain that is unable to adhere to cellulose and comparing the composition of proteins on the wild-type bacterium's cell membrane to those found on the membrane of the mutant strain. Using biological techniques, we were able to identify and characterize the proteins involved in the adhesion process.

We discovered that the digestive mechanism used by anaerobic bacteria like *Clostridium thermocellum* differs from that used by the aerobic (oxygen-requiring) fungus *Trichoderma reesei*. The fungus secretes a mixture of individual free proteins that break down cellulose and thus deconstruct the plant cell wall. By contrast, in some cellulose-degrading bacteria, *Clostridium thermocellum* among them, the enzymes are integrated into a larger enzymatic complex.[4] In the case of *Clostridium thermocellum*, the component that adheres to the cellulose is a large and discrete conglomerate that contains at least fourteen components, including an organizing protein and a large set of enzymes.

In the wake of our 1983 discovery, we decided to give this organelle the name that has since become standard: the cellulosome.[5] The cellulosome is formally defined today as a discrete complex of enzymes that includes various cellulose-degrading and associated enzymes dedicated to the degradation of the plant cell wall. Since that time many other strains of bacteria have been discovered that bear cellulosomes on their cell surfaces.

To grasp the way the cellulosome functions, one might think of it as a kind of Swiss army knife. Instead of various types of blades and other tools, the cellulosome has cellulose-degrading enzymes (and additional enzymes that hydrolyze other plant wall polysaccharides) of various types, all of which are attached to an organizing protein by means of a very strong biological recognition apparatus. When the cellulosome is in a dormant state, it is contained in protuberance-like structures on the bacterium's cell surface, but when it comes into contact with cellulose, the protuberances unravel and project extensions toward the cellulose, thereby attaching themselves to it via the cellulosomes. Each extension is equipped with a collection of enzymatic "blades" that immediately go into action.

How efficient is the cellulosome system? We compared the rate of cellulose degradation by the cellulosome with that of a collection of individual enzymes like that used by the fungus from Burma. We found that the sum total of the fungus's mixture of enzymes is much less efficient than the collection of enzymes contained in the cellulosome assembly.

4 Raphael Lamed, Eva Setter, & Edward A. Bayer, "Characterization of a cellulose-binding, cellulase-containing complex in *Clostridium thermocellum*," *Journal of Bacteriology*, 156 (1983), pp. 828–836.

5 Raphael Lamed, Eva Setter, Rina Kenig, & Edward A. Bayer, "The Cellulosome: A discrete cell surface organelle of *Clostridium thermocellum* which exhibits separate antigenic, cellulose-binding and various cellulolytic activities," *Biotechnology and Bioengineering Symposium*, 13 (1983), pp. 163–181.

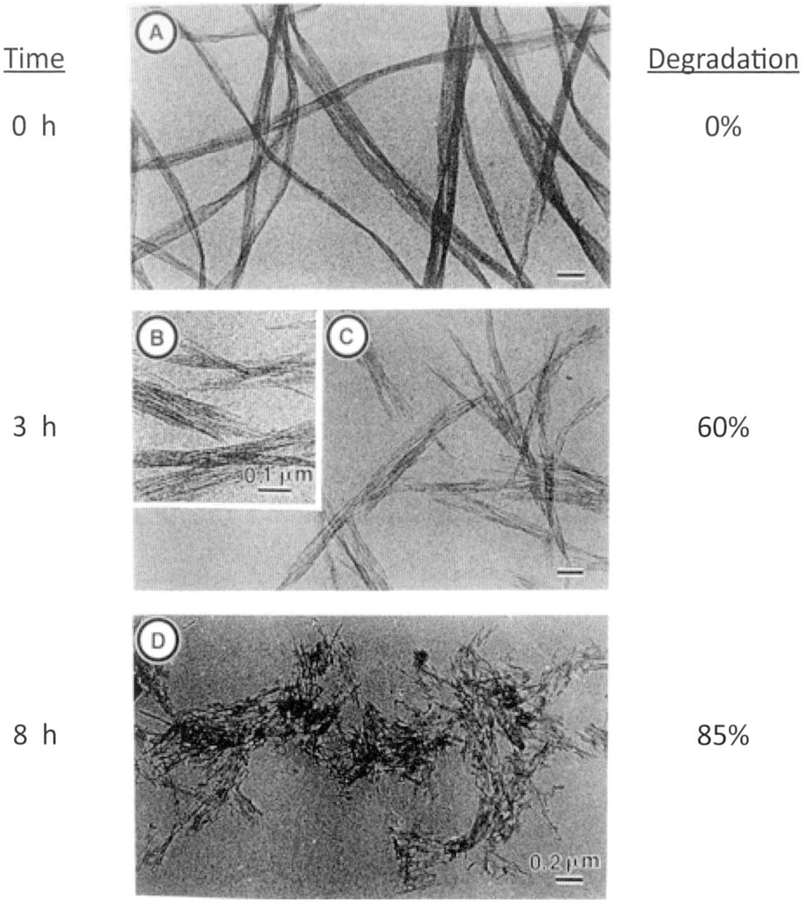

Figure 3: Transmission electron micrographs of crystalline cellulose prior to the cellulosome's action (A), three hours afterward (B, C), and eight hours later (D).

The accepted wisdom today is that both elements – the sophisticated and well-organized complex of cellulosome proteins, and the organelle's ability to adhere to cellulose – are necessary for efficient degradation of cellulose in general, and by anaerobic bacteria in particular.

What is the cellulosome's structure? Intensive study of the structure of the cellulosome conglomerate enabled us to precisely identify its component proteins, their modes of action, and the connections among them, between them and cellulose, and between them and the bacterium. We then characterized the structural components of specific function in the cellulosome and the way they work. We were able, for example, to determine the structure of the domain that attaches the cellulosome to cellulose and the molecular mechanism by which it binds to the carbohydrate. This research was conducted in collaboration with the late Jose Tormo of Yale University.[6]

The secret of how the cellulosome works was discovered to depend on its precise steering toward and binding to its target and on its delivery of the different enzymes in proximity to each other and the cellulose. This integrated process enables it to dismantle the cellulose and degrade it into simpler sugars, notably the two units of glucose called cellobiose. The uptake of these products by the bacterium, which also adheres to the cellulose by means of the cellulosome assembly, enhances the efficiency of the degradation.

Engineering a Synthetic "Super-Cellulosome"

These and other studies provided us with the level of understanding necessary to turn that basic knowledge into applied technology. We used the knowledge we had acquired to construct synthetic "designer" cellulosomes.[7] The potential of synthetic cellulosomes has been recognized by the United States Department of Energy during the present energy crisis.

Natural cellulosomes contain a more or less random mixture of cellulose-degrading enzymes. These random compositions provide for the bacterium's needs, but the industrial production of biofuels demands

6 Jose Tormo, Raphael Lamed, Arthur J. Chirino, Ely Morag, Edward A. Bayer, Yuval Shoham, & Thomas A. Steitz, "Crystal Structure of a Bacterial Family-III Cellulose-Binding Domain: A General Mechanism for Attachment to Cellulose," *EMBO Journal*, 15 (1996), pp. 5739–5751.

7 Edward A. Bayer, Ely Morag, and Raphael Lamed, "The Cellulosome: A Treasure Trove for Biotechnology," *Trends in Biotechnology*, 12 (1994), pp. 379–386.

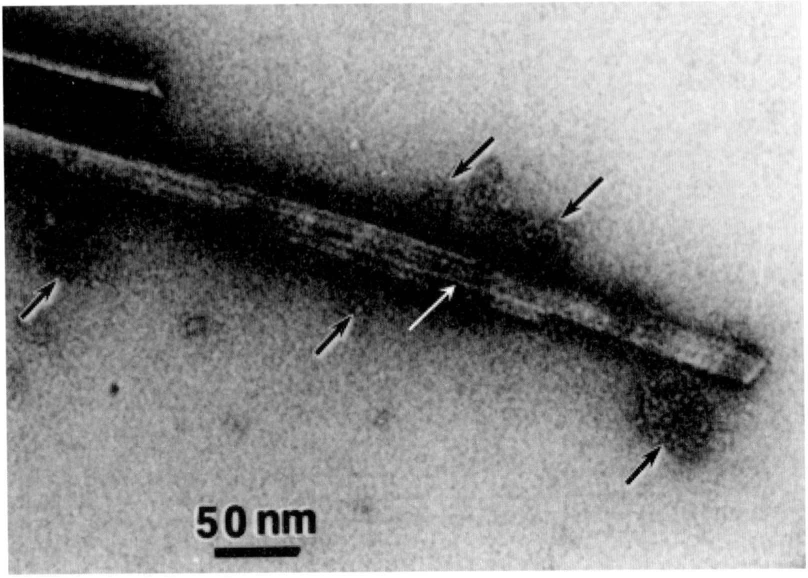

Figure 4: Cellulosomes (black arrows) attacking cellulose (white arrow). The particles (molecules) of the different enzymes can be seen, as well as the cellulosome's characteristic particulate structure and the fibrous nature of the cellulose.

maximal efficiency. To find the best cellulosome structure and composition for effectively breaking down the variety of cell walls presented by different kinds of plant cells – in other words, to create tailored artificial designer cellulosomes – we had to engineer cellulosome structures and study their characteristics. This was the next stage in our work.

We designed synthetic cellulosomes comprising different subunits containing the various cellulose-degrading and associated enzymes, which we placed in specific positions within the synthetic designer cellulosome. In this way we were able to construct an efficient cellulosome system containing six cellulose-degrading enzymes derived from different bacteria. The system we constructed achieves similar levels of activity to those of natural cellulosomes. The synthetic designer cellulosome is able to convert shredded paper into a mixture of soluble sugars in a single day.

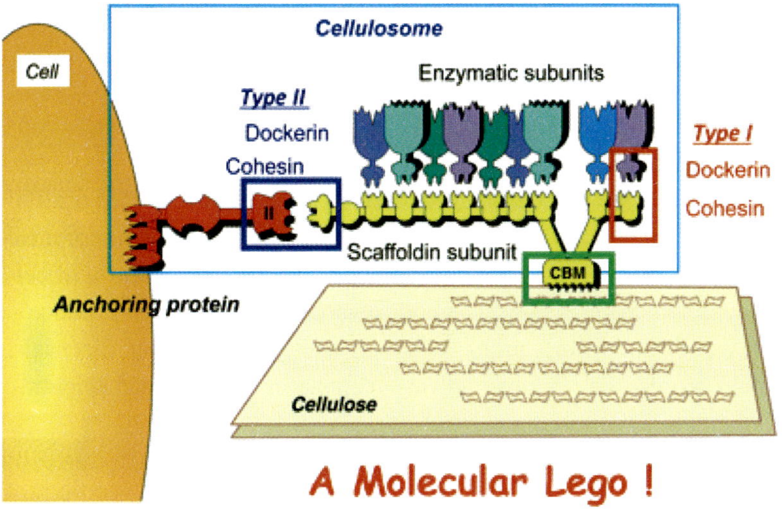

Figure 5: The cellulosome's organizing protein, scaffoldin, ties together the different enzymes by means of designated functional domains or modules found on the scaffoldin and complementary modular components of the enzymes (red rectangle). There is also a designated module that attaches the cellulosome to cellulose (green rectangle), as well as a modular pair that ties the cellulosome to the bacterium (blue rectangle), thereby attaching the entire bacterium to the cellulose.

As soon as we achieve the desired level of efficiency, we will continue to the next stage: the construction of bioreactors that can break down plant material, using engineered organisms such as bacteria, fungi, and yeast, into which we will have inserted the genes required for expression of the artificial cellulosomes needed to break down cellulose in a cost-effective manner.

From Imagination into Action

Surprisingly, it has recently been discovered that Weizmann's industrial bacterium is naturally able to produce an inefficient cellulosome, which requires improvement to enable it to degrade cellulose in a satisfactory way. We have thus come full circle. In the not-so-distant future, so we hope, what began as an idea of Weizmann's will ultimately be transformed into an efficient method of manufacturing biological fuels. The raw material will be cellulose, a problematic industrial byproduct. As noted above, cellulose is a particularly abundant renewable organic material that does not need to be produced at the expense of food crops such as corn.

Back in the 1940s, Weizmann wrote in a recently-discovered draft of his autobiography that the only research project of practical importance that he was still interested in completing was "the fermentation of cellulose, straw, wood, sawdust, wood residues, etc.," so as to produce – from wood, not from starch – alcohol, acetone, and other materials. In other words, our research reflects Weizmann's last applied scientific testament.[8]

In retrospect, Weizmann's vision can be considered the source of many of our ideas. In fact, we are using the strategy laid out by him many years ago. Incredibly, we, too, were advised at the outset of our research to "pour it down the sink." We hope that our ideas will soon become a reality, and that we will be able to use the bacteria we have studied and the knowledge we have acquired about cellulosomes to produce biofuels. The search for a solution to the global energy problem, neglected after the energy crisis of the 1970s, has made Chaim Weizmann's achievements more relevant than ever.

8 [See Section IV in the chapter "Unpublished Passages" in this volume. – Ed.]

Chaim Weizmann
List of Scientific Publications

compiled by Benjamin Z. Kedar

Abbreviations

Ber.	*Berichte der deutschen chemischen Gesellschaft*
Compt. rend.	*Comptes rendus hebdomadaires des séances de l'Académie des Sciences*
Ind. Eng. Chem.	*Industrial & Engineering Chemistry*
J. Amer. Chem. Soc.	*Journal of the American Chemical Society*
J. Chem. Soc.	*Journal of the Chemical Society* (until 1925: *Journal of the Chemical Society: Transactions*)
J. Org. Chem.	*Journal of Organic Chemistry*
J. Soc. Chem. Ind.	*Journal of the Society of Chemical Industry*
Proc. Chem. Soc.	*Proceedings of the Chemical Society, London*

1899

1 *I. Elektrolytische Reduktion von 1-Nitroanthrachinon. II. Ueber die Kondensation von Phenanthrenchinon u. 1-Nitroanthrachinon mit einigen Phenolen.*
Bern: Graphische Druckerei Henri Boneff. 53 pp.
Ch. WEIZMANN, Inaugural-Dissertation, Universität Freiburg, Schweiz

1903

2 Studien und Synthesen in der Reihe des Naphtacenchinons [I]
Chr. DEICHLER and Ch. WEIZMANN
Ber., 1903, **36**, 547–560

3 Studien und Synthesen in der Reihe des Naphtacenchinons [II]
Chr. DEICHLER and Ch. WEIZMANN
Ber., 1903, **36**, 719–728

4 Studien und Synthesen in der Reihe des Naphtacenchinons [III]
Chr. DEICHLER and Ch. WEIZMANN
Ber., 1903, **36**, 2326–2330

1904

5 The Effect of Anhydrides on Organo-Magnesium Bromides. Part I: The Action of Phthalic Anhydride on Magnesium-α-Naphthylbromide
Samuel Shrowder PICKLES and Charles WEIZMANN
Proc. Chem. Soc., 1904, **20**, 201

6 The Halogen Derivatives of Naphthacenchinone [*sic*]
Samuel Shrowder PICKLES and Charles WEIZMANN
Proc. Chem. Soc., 1904, **20**, 220

1905/6

7 Some Derivatives of Naphthoylbenzoic Acid and of Naphthacenequinone
Jan Quiller ORCHARDSON and Charles WEIZMANN
Proc. Chem. Soc., 1905, **21**, 307; *J. Chem. Soc.*, 1906, **89**, 115–121

8 Ethyl β-Naphthoylacetate
Charles WEIZMANN and Ernest Basil FALKNER
Proc. Chem. Soc., 1905, **21**, 307; *J. Chem. Soc.*, 1906, **89**, 122–125

9 Some Derivatives of Catechol, Pyrogallol, Benzophenone and of Substances Allied to the Natural Colouring Matters
William Henry PERKIN Jr. and Carl WEIZMANN [co-workers: Norman Allen Creeth, Victor John Harding, Walter Norman Haworth, Jonathan Naylor and Henry Llewellyn Smith]
Proc. Chem. Soc., 1906, **22**, 269; *J. Chem. Soc.*, 1906, **89**, 1649–1665

1906/7

10 4-Hydroxyphthalic and 4-Methoxyphthalic Acids
William Henry BENTLEY and Charles WEIZMANN
Proc. Chem. Soc., 1906, **22**, 323; *J. Chem. Soc.*, 1907, **91**, 98–104

11 3-Hydroxyphthalic and 3-Methoxyphthalic Acids and Their Derivatives
William Henry BENTLEY, Rona ROBINSON and Charles WEIZMANN
Proc. Chem. Soc., 1906, **22**, 323; *J. Chem. Soc.*, 1907, **91**, 104–112

12 Derivatives of Naphthacenequinone
William Henry BENTLEY, Arthur FRIEDL, Frederick THOMAS and Charles WEIZMANN [co-workers: E.C.C. Baly and W.B. Tuck]
Proc. Chem. Soc., 1906, **22**, 324; *J. Chem. Soc.*, 1907, **91**, 411–435

13 The Fluoresceins and Eosins from 4-Hydroxyphthalic, 4-Methoxyphthalic and Hemipinic Acids
Arthur FRIEDL, Charles WEIZMANN and Max WYLER
Proc. Chem. Soc., 1907, **23**, 214; *J. Chem. Soc.*, 1907, **91**, 1584–1588

14 Derivatives of Naphthacenequinones [*sic*]. Part II
William Henry BENTLEY, Arthur FRIEDL and Charles WEIZMANN
Proc. Chem. Soc., 1907, **23**, 215; *J. Chem. Soc.*, 1907, **91**, 1588–1593

15 Researches on Anthraquinones and Phthaleins
William Henry BENTLEY, Henry Dent GARDNER Jr. and Charles WEIZMANN
Proc. Chem. Soc., 1907, **23**, 215; *J. Chem. Soc.*, 1907, **91**, 1626–1640

1908

16 Researches on the Anthraquinones
William Henry BENTLEY and Charles WEIZMANN
Proc. Chem. Soc., 1908, **24**, 52; *J. Chem. Soc.*, 1908, **93**, 435–438

1909

17 Derivatives of Naphthacenequinone. Part III
Dorothy HARROP, Roland Victor NORRIS and Charles WEIZMANN
J. Chem. Soc., 1909, **95**, 279–288

18 Some Derivatives of Anthraquinone
Dorothy HARROP, Roland Victor NORRIS and Charles WEIZMANN
Proc. Chem. Soc., 1909, **25**, 203; *J. Chem. Soc.*, 1909, **95**, 1312–1319

1910/11

19 Δ[1]-Nonylenic Acid
Victor John HARDING and Charles WEIZMANN
Proc. Chem. Soc., 1910, **26**, 24; *J. Chem. Soc.*, 1910, **97**, 299–304

20 1:4-Dichloanthraquinone and Its Derivatives
Gertrude Maud WALSH and Charles WEIZMANN
Proc. Chem. Soc., 1910, **26**, 61; *J. Chem. Soc.*, 1910, **97**, 685–692

21 Synthesis of Dipeptides of Lauric and *n*-Nonoic Acids. Products of the Condensation of Lauric and *n*-Nonoic Acids with Glycine, Alanine and Leucine
Arthur HOPWOOD and Charles WEIZMANN
Proc. Chem. Soc., 1910, **26**, 69

22 Synthesis of 6-Carboxy-3:4-Dimethoxyphenylglyoxolic Acid
Victor John HARDING and Charles WEIZMANN
Proc. Chem. Soc., 1910, **26**, 130; *J. Chem. Soc.*, 1910, **97**, 1126–1131

23 Attempts to Prepare Glycerides of Amino-acids
Roman ALPERN and Charles WEIZMANN
Proc. Chem. Soc., 1910, **26**, 345; *J. Chem. Soc.*, 1911, **99**, 84–87

24 β-Methyl-$\Delta^{\alpha\lambda}$-dodecadiene and β-Methyl-$\Delta^{\alpha\gamma}$-decadiene
Victor John HARDING, Gertrude Maud WALSH and Charles WEIZMANN
Proc. Chem. Soc., 1911, **27**, 12; *J. Chem. Soc.*, 1911, **99**, 448–451

25 Synthesis of Dipeptides of α-Aminolauric Acid with Glycine, Alanine, Valine, Leucine and Asparagine
Arthur HOPWOOD and Charles WEIZMANN
Proc. Chem. Soc., 1911, **27**, 55; *J. Chem. Soc.*, 1911, **99**, 571–576

26 Synthesis of Polypeptides of α-Amino-*n*-Nonoic Acid with Glycine, Alanine, Valine, Leucine, Asparagine and Aspartic Acid
Arthur HOPWOOD and Charles WEIZMANN
Proc. Chem. Soc., 1911, **27**, 214; *J. Chem. Soc.*, 1911, **99**, 1577–1585

1912

27 αδ-Derivatives of Adipic and β-Methyladipic Acids
Harold DAVIES, Henry STEPHEN and Charles WEIZMANN
Proc. Chem. Soc., 1912, **28**, 94–95

28 Condensation of Acid Chlorides with the Ethyl Esters of (a) Cyanoacetic Acid, (b) Malonic Acid and (c) Acetoacetic Acid. Preliminary Note.
Charles WEIZMANN, Harold DAVIES and Henry STEPHEN
Proc. Chem. Soc., 1912, **28**, 103–104

29 Tyrosine and its Derivatives Containing Substituents in the Benzene Ring
Henry STEPHEN and Charles WEIZMANN
Proc. Chem. Soc., 1912, **28**, 160

30 Condensation of Bromoacyl Haloids with Glucosamine
Arthur HOPWOOD and Charles WEIZMANN
Proc. Chem. Soc., 1912, **28**, 261–262

1913

31 αδ-Derivatives of Adipic and β-Methyladipic Acids and the Preparation of Muconic and β-Methylmuconic Acids
Henry STEPHEN and Charles WEIZMANN
Proc. Chem. Soc., 1913, **29**, 14; *J. Chem. Soc.*, 1913, **103**, 269–276

32 Synthesis of the Anhydrides of α-Aminoacyl Glucosamines
Charles WEIZMANN and Arthur HOPWOOD
Proceedings of the Royal Society of London, Series A, 1913, **88**, 455–461

33 The Preparation of Some Organo-Selenium Compounds
Charles WEIZMANN and Henry STEPHEN
Proc. Chem. Soc., 1913, **29**, 196

34 Condensation of Acid Chlorides with the Ethyl Esters of (a) Cyanoacetic Acid, (b) Malonic Acid, (c) Acetoacetic Acid. Part I
Charles WEIZMANN, Henry STEPHEN and Ganesh Sakharam AGASHE
Proc. Chem. Soc., 1913, **29**, 261; *J. Chem. Soc.*, 1913, **103**, 1855–1861

35 Hydrolysis of Proteins with an Alcoholic Solution of Hydrogen Chloride. Part I
Charles WEIZMANN and Ganesh Sakharam AGASHE
Biochemical Journal, 1913, **7**, 437–440

1914

36 Derivatives of 3:4-Dimethoxyacetophenone and 4:5-Dimethoxy-*o*-Tolyl Methyl Ketone, and the Synthesis of Phenylglyoxalines Containing Substituents in the Benzene Ring
Henry STEPHEN and Charles WEIZMANN
Proc. Chem. Soc., 1914, **30**, 71; *J. Chem. Soc.*, 1914, **105**, 1046–1057

37 Synthesis of *d,l*-Tyrosine and *d,l*-3,4-Dihydroxyphenylalanine
Henry STEPHEN and Charles WEIZMANN
Proc. Chem. Soc., 1914, **30**, 114; *J. Chem. Soc.*, 1914, **105**, 1152–1155

38 Some Homologues of Alizarin
Harry BRADBURY and Charles WEIZMANN
J. Chem. Soc., 1914, **105**, 2748–2751

1915

39 Condensation of Acid Chlorides with the Ethyl Esters of (a) Cyanoacetic Acid, (b) Malonic Acid, and (c) Acetoacetic Acid. Part II: Some Experiments on Ethyl Phthaliminoacetoacetate and Ethyl γ-Ethoxyacetoacetate
John BRADSHAW, Henry STEPHEN and Charles WEIZMANN
J. Chem. Soc., 1915, **107**, 803–813

40 Phthalides of the Benzene, Naphthalene and Carbazole Series. Part I
Maurice COPISAROW and Charles WEIZMANN
J. Chem. Soc., 1915, **107**, 878–886

1920

41 Some Condensations of *n*-Butyl Alcohol and *n*-Butaldehyde
Charles WEIZMANN and Stanley Frederick GARRARD
J. Chem. Soc., 1920, **117**, 324–338

1934

42 Actions des rayons ultraviolets sur le glycocolle
V. HENRI, Ch. WEIZMANN and Y. HIRSHBERG
Compt. rend., 1934, **198**, 168–170

43 Décomposition photochimique du glycocolle. Influence du milieu et de la longueur d'onde
V. HENRI, Ch. WEIZMANN and Y. HIRSHBERG
Compt. rend., 1934, **199**, 1205–1207

1935

44 Grignard Reactions with Phthalic Anhydrides
Chaim WEIZMANN, Ernst BERGMANN and Felix BERGMANN
J. Chem. Soc., 1935, 1367–1370

45 The Reaction of Grignard Reagents with Some Succinic Anhydrides
Chaim WEIZMANN, Ottilie BLUM-BERGMANN and Felix BERGMANN
J. Chem. Soc., 1935, 1370–1371

1936

46 Grignard Reactions with Ethyl β-Chloropropionate. Part I
Ch. WEIZMANN and Ernst BERGMANN
J. Chem. Soc., 1936, 401–402

47 Methoxylated *o*-Benzoylbenzoic Acids
Ch. WEIZMANN and Ernst BERGMANN
J. Chem. Soc., 1936, 567–569

48 The Photochemical Deamination of Amino Acids in Water Solution. Part I
Ch. WEIZMANN, E. BERGMANN and Y. HIRSHBERG
J. Amer. Chem. Soc., 1936, **58**, 1675–1678

1937

49 Mechanism and Applicability of the Guerbet Reaction
Ch. WEIZMANN, Ernst BERGMANN and L. HASKELBERG
J. Soc. Chem. Ind., 1937, **56**, 587–591

50 Note on the Butyl Alcohol-Acetone Fermentation of Maize
Charles WEIZMANN and Harold DAVIES
J. Soc. Chem. Ind., 1937, **56**, 8–10 T

51 The Activation of the Butanol-Acetone Fermentation of Carbohydrates by *Clostridium acetobutylicum* (Weizmann)
Chaim WEIZMANN and Bruno ROSENFELD
Biochemical Journal, 1937, **31**, 619–639

52 Über die Aceton-Butylalkohol-Gärung
Ernst SIMON and Chaim WEIZMANN
Enzymologia, 1937, **4** [= Festschrift Carl Neuberg], 169–188

1938

53 Preliminary Note on an Ethyl-Alcohol-Producing Bacillus Isolated from *Hibiscus sabdariffa* L. (Rozella)
Charles WEIZMANN
J. Soc. Chem. Ind., 1938, **57**, 1049–1050

54 Synthesis of Polycyclic Compounds from Dicyclohexenyl
Ch. WEIZMANN, Ernst BERGMANN and T. BERLIN
J. Amer. Chem. Soc., 1938, **60**, 1331–1334

55 Photochemical Interaction between Ketones and Alcohols
Ch. WEIZMANN, Ernst BERGMANN and Y. HIRSHBERG
J. Amer. Chem. Soc., 1938, **60**, 1530–1533

56 Photochemical Interaction between Ketones and Secondary Alcohols
Ch. WEIZMANN, Y. HIRSHBERG and Ernst BERGMANN
Nature, 1938, **141**, 1012–1013

57 The Photochemical Deamination of Amino Acids in Water Solution. Part II
Ch. WEIZMANN, Y. HIRSHBERG and Ernst BERGMANN
J. Amer. Chem. Soc., 1938, **60**, 1799–1801

58 Interaction between Grignard Compounds and Maleic Acid Derivatives
Ch. WEIZMANN and F. BERGMANN
J. Amer. Chem. Soc., 1938, **60**, 2647–2650

59 Polycyclic Aromatic Hydrocarbons
Ch. WEIZMANN and Ernst BERGMANN
Scripta Academica Hierosolymitana. Scientific Report, 1 (Jerusalem 1938), 1–32 (with Hebrew summary)

1939

60 Phenylated Phthalic Acids and Anthracene Derivatives
Ch. WEIZMANN, Ernst BERGMANN and L. HASKELBERG
J. Chem. Soc., 1939, 391–397

61 The Boric Acid Synthesis of *peri*-Hydroxyanthraquinones
Ch. WEIZMANN, L. HASKELBERG and T. BERLIN
J. Chem. Soc., 1939, 398–401

62 1:4:9:10-Tetraphenylanthracene
Ch. WEIZMANN and Ernst BERGMANN
J. Chem. Soc., 1939, 494–495

63 Hydrolytic Deamination of Amino Acids
Ch. WEIZMANN, Ernst BERGMANN and Y. HIRSHBERG
Nature, 1939, **143**, 723–724

64 The Specific Nutritive Requirements of *Clostridium acetobutylicum* (Weizmann). Part II
Ch. WEIZMANN and Bruno ROSENFELD
Biochemical Journal, 1939, **33**, 1376–1389

1940

65 Identification and Determination of Aromatic Compounds in Mineral Oils
Ch. WEIZMANN, Victor HENRI and Ernst BERGMANN
Nature, 1940, **146**, 230–231

66 Studies on some Strains of Butyric Acid-Producing Plectridia Isolated from Hemp, Jute and Flax
Ch. WEIZMANN and Esther HELLINGER
Journal of Bacteriology, 1940, **40**, 665–682

67 Some Syntheses of Biochemical Interest
Chaim WEIZMANN and Ernst BERGMANN
Medical Leaves: A Review of the Jewish Medical World and Medical History, 1940, **3**, 223–230

1944

68 Attempted Synthesis of Hemipinic Acid from Guaiacol
Ch. WEIZMANN and L. HASKELBERG
J. Org. Chem., 1944, **9**, 121–124

69 Studies on Certain Butyric Acid-Producing Organisms Isolated from Hemp and Similar Vegetable Fibres
Ch. WEIZMANN and Esther HELLINGER
Palestine Journal of Botany, Rehovot Series, 1944, **4/2**, 51–76

1947

70 Carbohydrates as an Alternative Source to Petroleum for Organic Chemicals
Ch. WEIZMANN, Ernst BERGMANN and M. SULZBACHER
Chemistry and Industry, 1947, **50**, 769–771

71 α-CycloHexil-β-propylacrilic Acid from Cyclohexanol
Ch. WEIZMANN, M. SULZBACHER and E. BERGMANN
J. Chem. Soc., 1947, 772–773

72 Reduction of α,β-Unsaturated Carbonyl Compounds by Sodium Butoxide
Ch. WEIZMANN, M. SULZBACHER and E. BERGMANN
J. Chem. Soc., 1947, 851

1948

73 Production of Aromatic Hydrocarbons from Petroleum
Ch. WEIZMANN, E. BERGMANN, H. STEINER and M. SULZBACHER
J. Soc. Chem. Ind., 1948, **67**, 114–118

74 Studies in Selective Extraction and Adsorption. I. Selective Solvents for Ethylene Chlorohydrin
Ch. WEIZMANN, E. BERGMANN, E.F. CHANDLEY, H. STEINER, M. SULZBACHER and E. ZIMKIN
J. Soc. Chem. Ind., 1948, **67**, 203–205

75 Studies in Selective Extraction and Adsorption. III. The Adsorption of Acetone, Butyl Alcohol and 2:3-Butylene Glycol from Dilute Solutions
Ch. WEIZMANN, E. BERGMANN, M. SULZBACHER and E.R. PARISER
J. Soc. Chem. Ind., 1948, **67**, 225–227

76 The Synthesis of α-Alkoxyisobutyric Acids and Alkyl Methacrylates from Acetonechloroform
Ch. WEIZMANN, M. SULZBACHER and E. BERGMANN
J. Amer. Chem. Soc., 1948, **70**, 1153–1158

77 Preparation of Acetonechloroform and its Homologs
Ch. WEIZMANN, E. BERGMANN and M. SULZBACHER
J. Amer. Chem. Soc., 1948, **70**, 1189–1191

78 Derivatives of Salicylic Acid
Ch. WEIZMANN, Ernst D. BERGMANN and M. SULZBACHER
J. Org. Chem., 1948, **13**, 796–799

1950

79 Further Observations on the Guerbet Reaction
Ch. WEIZMANN, Ernst BERGMANN and Max SULZBACHER
J. Org. Chem., 1950, **15**, 54–57

80 The Use, for Condensation Reactions, of Potassium Hydroxide in Solvents of the Acetal Type
Ch. WEIZMANN, E. BERGMANN and M. SULZBACHER
J. Org. Chem., 1950, **15**, 918–929

1951

81 Aromatizing Cracking of Hydrocarbon Oils. [I]. Process Data and Mechanism
Chaim WEIZMANN, Ernst BERGMANN, W.E. HUGGETT, Herbert STEINER and Max SULZBACHER [with Donald Parker, K.O. Michaelis, Sydney Whincup and Emanuel Zimkin]
Ind. Eng. Chem., 1951, **43**, 2312–2318

82 Aromatizing Cracking of Hydrocarbon Oils. [II]. Polycyclic Fractions.
Chaim WEIZMANN, Ernst BERGMANN, H.S. BOYD-BARRETT, Herbert STEINER and Max SULZBACHER [with J.R. Holker, Erna Mandel, John Porges and Derrick Rowley]
Ind. Eng. Chem., 1951, **43**, 2318–2321

83 Aromatizing Cracking of Hydrocarbon Oils. [III]. Influence of the Nature of the Charging Stock
Chaim WEIZMANN, Ernst BERGMANN, W.E. HUGGET, Herbert STEINER, Max SULZBACHER, K.O. MICHAELIS, Felix POPPER, Sydney WHINCUP and Emanuel ZIMKIN
Ind. Eng. Chem., 1951, **43**, 2322–2325

84 Aromatizing Cracking of Hydrocarbon Oils. [IV]. Conversion of Butylene into Aromatics
Chaim WEIZMANN, Victor HENRI and Ernst BERGMANN
Ind. Eng. Chem., 1951, **43**, 2325–2326

Chaim Weizmann

List of Patents

compiled by Yigal Burstein

1 Verfahren zur Darstellung von Naphtacenchinonderivaten.
Chr. DEICHLER and Ch. WEIZMANN
D.R.P. 134,985 (D.10969) K1. 12q. Sept. 14, 1900; F.P. 320,327, Apr. 10, 1902.

2 Process of manufacturing dihydroxynaphthacenquinones.
Chr. DEICHLER and Ch. WEIZMANN
D.R.P. 138,324 (D.11226) K1. 12q. Sept. 14, 1900; B.P. 9,038, Apr. 18, 1902.

3 Verfahren zur Darstellung von Naphtacenchinonderivaten.
Chr. DEICHLER and Ch. WEIZMANN
D.R.P. 138,325 (D.11579) K1. 12q. Sept. 14, 1900.

4 Verfahren zur Darstellung von *o*-Oxynaphthoylbenzoesäure.
Chr. DEICHLER and Ch. WEIZMANN
D.R.P. 141,025 (D.11135) K1. 12 q. Nov. 30, 1900.

5 Process of manufacturing bromo-derivatives of naphthoylbenzoic acid from bromophthalic acids.
Charles WEIZMANN
U.S.P. 753,372, July 5, 1903.

6 Improvements in the manufacture of anthracene derivatives and dye stuffs therefrom.
Alexander MEYENBERG and Charles WEIZMANN
B.P. 190508744, Apr. 25, 1905.

7 Process of manufacturing camphene from pinene-hydrochloride.
Charles WEIZMANN
B.P. 190605032, Mar. 1, 1906; F.P. 369,449, Sept. 4, 1906; D.R.P. 197,163, Sept. 8, 1906; U.S.P. 896,962.

8 Improvements in the manufacture of isobornyl-esters.
Charles WEIZMANN
B.P. 190608266, Apr. 5, 1906.

9 Improvements in the manufacture of isobornyl esters.
Charles WEIZMANN
B.P. 190610798, May 8, 1906.

10 Improvements in the manufacture of isobornyl esters.
Charles WEIZMANN
B.P. 1906018280, Aug. 14, 1906.

11 Improvements in the treatment of turpentine for use in the manufacture of pinene hydrochloride.
Charles WEIZMANN
B.P. 190623875, Oct. 26, 1906.

12 Process of purifying turpentine.
Charles WEIZMANN
B.P. 190623878, Oct. 26, 1906.

13 Process of obtaining fatty acids isobornyl esters.
Charles WEIZMANN
D.R.P. Anm. W. 27,218, Feb. 15, 1907; F.P. 375,007, Feb. 19, 1907; F.P. 379,687, July 6, 1907; D.R.P. 207,155, Feb. 16, 1907.

14 Manufacture of a mixture of camphene isobornylacetate.
Ch. WEIZMANN
D.R.P. 205,849, July 2, 1907.

15 Improvements in the manufacture of isobornyl esters.
Charles WEIZMANN
B.P. 190716605, July 19, 1907; U.S.P. 910,978, Mar. 5, 1908.

16 Improvements in the manufacture of camphor.
Arthur FRIEDL and Charles WEIZMANN
B.P. 190721946, Oct. 4, 1907.

17 Improvements in the treatment of liquid, or oily, pinene-hydrochloride, and the obtainment of solid pinene-hydrochloride and camphene therefrom.
Charles WEIZMANN and William Henry BENTLEY
B.P. 190807319, Apr. 2, 1908.

18 Improvements connected with the synthetic manufacture of caoutchouc and an intermediate product.
William Henry PERKIN, Charles WEIZMANN, Francis Edward MATTHEWS and Edward Halford STRANGE
B.P. 19105932, Mar. 9, 1910; F.P. 427,286, Mar. 7, 1911.

19 Process of manufacturing isoprene.
William Henry PERKIN and Charles WEIZMANN
U.S.P. 991,453, Mar. 17, 1911.

20 Improvements connected with the synthetic manufacture of caoutchouc-like bodies and intermediate products.
William Henry PERKIN, Charles WEIZMANN, Francis Edward MATTHEWS and Edward Halford STRANGE
B.P. 191109722, Apr. 21, 1911.

21 Manufacture of halogen derivatives of organic compounds.
William Henry PERKIN, Charles WEIZMANN and Harold DAVIES
B.P. 191200277, Jan. 3, 1912; U.S.P. 1,144,237, Dec. 30, 1912; F.P. 452,593, Dec. 28, 1912.

22 Halogen substituted organic compounds.
William Henry PERKIN and Charles WEIZMANN
B.P. 191222653, Oct. 4, 1912; Can. P. 153,266, Dec. 31, 1912 (with Harold DAVIS)

23 Improvements in the Bacterial Fermentation of Carbohydrates and in Bacterial Cultures for the Same.
Charles WEIZMANN
B.P. 191504845, Mar. 29, 1915; U.S.P. 1,437,697, May 1, 1918; D.R.P. 445,982, July 8, 1920; Austrian P. 95,449, July 26, 1920.

24 Improvements relating to fermentation processes for the production of acetone and butyl alcohol.
Charles WEIZMANN
B.P. 149,355, Feb. 24, 1916.

25 Improvements relating to the manufacture of secondary butyl alcohol.
Charles WEIZMANN and David Alliston LEGG
B.P. 161,591, Feb. 24, 1916; U.S.P. 1,408,320, Nov. 10, 1917; Can. P. 202,135, July 20, 1920.

26 Improvements relating to the manufacture of hydrocarbons.
Charles WEIZMANN and David Alliston LEGG
B.P. 165,452, Mar. 7, 1916; U.S.P. 1,395,620, Oct. 11, 1917; Can. P. 203,408, Aug. 31, 1920.

27 Improvements relating to fermentation processes for the production of acetone and butyl alcohol.
Charles WEIZMANN and George Anthony HAMLYN
B.P. 164,023, June 29, 1916; U.S.P. 1,329,214, Nov. 27, 1918.

28 Production of acetone and alcohol by bacteriological processes.
Charles WEIZMANN
U.S.P. 1,315,585, Dec. 26, 1916.

29 Fermentation process for the production of acetone and butyl alcohol.
Charles WEIZMANN
B.P. 150,360, May 25, 1917; U.S.P. 1,437,697, May 1, 1918.

30 Glycols.
Charles WEIZMANN and Ernest Graham BAINBRIDGE
B.P. 131,628, June 3, 1918.

31 Improvements relating to the fermentation of cereals and other starchy materials.
Charles WEIZMANN and Henry Michael SPIERS
B.P. 164,366, June 3, 1918.

32 Improvements relating to the fermentation of starchy material for the production of acetone and butyl alcohol.
Charles WEIZMANN and Henry Michael SPIERS
B.P. 164,762, June 3, 1918.

33 Ethyl methyl ketone.
Charles WEIZMANN
Can. P. 203,264, Aug. 24, 1920.

34 Manufacture of hydrocarbons.
Charles WEIZMANN and David Alliston LEGG
Can. P. 203,408, Aug. 31, 1920.

35 Improvements relating to the production of titanic acid.
Charles WEIZMANN and Joseph BLUMENFELD
B.P. 203,352, Mar. 1, 1922.

36 Improvements relating to titanium compounds.
Charles WEIZMANN and Joseph BLUMENFELD
B.P. 209,441, July 12, 1922.

37 Improvements relating to titanium compounds.
Charles WEIZMANN and Joseph BLUMENFELD
B.P. 209,480, July 12, 1922.

38 Improvements relating to titanium compounds.
Charles WEIZMANN and Joseph BLUMENFELD
B.P. 210,033, July 12, 1922.

39 Improvements relating to titanium pigments.
Charles WEIZMANN and Joseph BLUMENFELD
B.P. 247,296, July 12, 1922; F.P. 575,615; Can. P. 263,084.

40 Treating titanium ore.
Charles WEIZMANN and Joseph BLUMENFELD
U.S.P. 1,489,183, May 17, 1923.

41 Improvements relating to the preparation of titanium hydroxide.
Joseph BLUMENFELD and Charles WEIZMANN
B.P. 225,593, July 31, 1923.

42 Improvements relating to filtration of solutions and the like.
Charles WEIZMANN and Joseph BLUMENFELD
B.P. 227,143, Aug. 3, 1923.

43 Improvements relating to the treatment of solutions for the separation of suspended matter.
Charles WEIZMANN and Joseph BLUMENFELD
B.P. 228,814, Aug. 3, 1923.

44 Improvements relating to the preparation of pigments and paints.
Charles WEIZMANN and Joseph BLUMENFELD
B.P. 256,302, Apr. 22, 1925.

45 Improvements relating to yeast or protein preparations.
Charles WEIZMANN
B.P. 450,529, Oct. 20, 1934.

46 Improvements relating to yeast or protein preparations.
Charles WEIZMANN
B.P. 478,570, July 16, 1936.

47 Improvements relating to foodstuffs.
Charles WEIZMANN
B.P. 483,587, Oct. 28, 1936; F.P. 837,651, Oct. 28, 1937.

48 Improvements relating to the production of fatty acids.
Charles WEIZMANN
B.P. 489,170, Jan. 18, 1937; F.P. 840,785.

49 Improvements in the production of butyl alcohol-acetone and butyl alcohol.
Charles WEIZMANN
B.P. 496,137, May 24, 1937; F.P. 838,258, Mar. 2, 1939.

50 Preparations of yeast or proteins.
Charles WEIZMANN
F.P. 834,322, July 19, 1937.

51 Improvements relating to protein preparations.
Charles WEIZMANN
B.P. 509,495, Oct. 15, 1937; F.P. 852,540, Feb. 26, 1940; U.S.P. 2,217,264, Oct. 8, 1940; Pal. P. 18,921, 1941; India P. 28,048, 1941.

52 Improvements in and relating to the production of acetone and butyl alcohol by fermentation.
Charles WEIZMANN
B.P. 518,962, Apr. 9, 1938: F.P. 852,298, Mar. 29, 1939.

53 Cracking hydrocarbons.
Victor HENRI, Charles WEIZMANN and Joseph BLUMENFELD
F.P. 849,456, July 29, 1938.

54 Process for the cracking of oils.
Alfred Claude JESSUP (to Charles WEIZMANN)
F.P. 854,041, Dec. 5, 1938; B.P. 536,891 (by Jessup), Nov. 16, 1939; U.S.P. 2,313,241, Mar. 9, 1943.

55 Process for the production of isobutylene and its dimer.
Alfred Claude JESSUP (to Charles WEIZMANN)
F.P. 854,042, Dec. 5, 1938; U.S.P. 2,303,362, Nov. 16, 1939; B.P. 536,734 (by Jessup), Nov. 30, 1939.

56 Process of degrading protein products.
Charles WEIZMANN
U.S.P. 2,141,455, Dec. 27, 1938.

57 Improvements relating to alimentary preparations.
Charles WEIZMANN
B.P. 529,775, May 16, 1939.

58 Improvements relating to alimentary preparations.
Charles WEIZMANN
B.P. 537,423, Nov. 20, 1939; U.S.P. 2,316,733, Jan. 21, 1941.

59 Improvements relating to yeast preparations.
Charles WEIZMANN
B.P. 516,343, Jan. 1, 1940; F.P. 859,591; U.S.P. 2,225,576, Dec. 17, 1940.

60 Improvements in and relating to the utilization of coal oil.
Charles WEIZMANN
B.P. 574,963, May 13, 1940.

61 Process of degrading protein products.
Charles WEIZMANN
U.S.P. 2,206,622, July 2, 1940.

62 Improvements in and relating to the manufacture of aromatic hydrocarbons.
Charles WEIZMANN
B.P. 552,115, July 12, 1940; Trinidad 15, 1942; Pal. P. 2,159, 1942.

63 Improvements in and relating to the production of aromatic hydrocarbons.
Charles WEIZMANN
B.P. 552,216, July 12, 1940.

64 Improvements in and relating to the manufacture of aromatics from gaseous olefins with oxides.
Charles WEIZMANN
B.P. 552,551, July 12, 1940; U.S.P. 2,329,672, Oct. 29, 1940; Trinidad 14, 1942; Pal. P. 2,158, 1942.

65 Improvements in and relating to processes for the production of acetone and butyl alcohol by fermentation.
Charles WEIZMANN
B.P. 572,637, Mar. 11, 1941.

66 Improvements in and relating to the production of aromatic hydrocarbons.
Charles WEIZMANN
B.P. 574,973, Mar. 11, 1941.

67 Process for the manufacture of unsaturated ketones.
Charles WEIZMANN (to Percy MAY)
B.P. 569,373, Apr. 11, 1941; U.S.P. 2,388,101, Nov. 6, 1942.

68 Improvements in and relating to the fermentation of molasses.
Charles WEIZMANN
B.P. 572,641, Apr. 11, 1941.

69 Improvements in and relating to the production of aliphatic acids by the fermentation of difficultly fermentable materials.
Charles WEIZMANN
B.P. 572,642, Apr. 11, 1941.

70 Improvements in and relating to fermentation processes.
Charles WEIZMANN
B.P. 572,643, Apr. 11, 1941; U.S.P. 2,417,801, Mar. 2, 1942 (to Butacet Ltd.).

71 Improvements in and relating to the utilisation of sulphite waste liquors, namely, by the production of aliphatic acids from sulfite-waste liquor.
Charles WEIZMANN
B.P. 572,644, Apr. 11, 1941; U.S.P. 2,415,777, Mar. 22, 1943 (to Butacet Ltd.).

72 Process for the manufacture of ketones.
Charles WEIZMANN
B.P. 572,645, Apr. 11, 1941.

73 Process for the production of aliphatic ketones.
Charles WEIZMANN
B.P. 573,724, Apr. 11, 1941.

74 Improvements in the fermentation production of a mixture of fatty acids.
Charles WEIZMANN
B.P. 573,930, Apr. 11, 1941.

75 Process for the production of saturated ketones.
Charles WEIZMANN
B.P. 574,446, Apr. 11, 1941.

76 Improvements in the treatment of olefinic hydrocarbons for the production of alcohols, ketones and other substances.
Charles WEIZMANN
B.P. 575,571, Apr. 17, 1941.

77 Process for the production of ketones.
Charles WEIZMANN
B.P. 550,099, Apr. 18, 1941.

78 Improvements relating to the production of carboxylic acids.
Charles WEIZMANN
B.P. 550,134, Apr. 18, 1941.

79 Improvements in the synthesis of secondary and tertiary ethynylcarbinols.
Charles WEIZMANN
B.P. 573,527, Apr. 18, 1941.

80 Improvements in the synthesis of secondary and tertiary ethynylcarbinols.
Charles WEIZMANN
B.P. 580,921, Apr. 18, 1941.

81 Improvements in and relating to the production of aromatic compounds.
Charles WEIZMANN
B.P. 575,766, Apr. 23, 1941.

82 Improvements in and relating to the production of aromatic bodies.
Charles WEIZMANN
B.P. 575,771, Apr. 23, 1941.

83 Improvements in and relating to fermentation processes.
Charles WEIZMANN
B.P. 572,762, May 24, 1941.

84 Improvements in and relating to the construction of cracking plants and the like.
Charles WEIZMANN and Herbert STEINER
B.P. 551,215, Aug. 6, 1941.

85 Improvements in and relating to the fermentation production of acetone and butyl alcohol.
Charles WEIZMANN
B.P. 572,763, Aug. 6, 1941.

86 Improvements in and relating to the carrying out of chemical processes and to apparatus therefor.
Charles WEIZMANN, Ernst BERGMANN, Herbert STEINER and Henry Ernest CHARLTON
B.P. 575,383, Aug. 6, 1941.

87 Improvements relating to motor fuels.
Charles WEIZMANN
B.P. 575,636, Aug. 11, 1941.

88 Improvements relating to motor fuels.
Charles WEIZMANN
B.P. 575,637, Aug. 11, 1941.

89 Process for the conversion of c.v.r. benzole into a substantially completely aromatic product.
Charles WEIZMANN
B.P. 575,768, Sept. 16, 1941.

90 Process for the production of aromatic compounds from solvent naphtha.
Charles WEIZMANN
B.P. 575,769, Sept. 16, 1941.

91 Proteolysis with fig juice.
Charles WEIZMANN
Pal. P. 1,891, 1941; India P. 28,048, 1941.

92 Production of aromatic bodies.
Charles WEIZMANN
U.S.P. 2,397,715, Jan. 9, 1942.

93 Improvements in and relating to the conversion of olefins into diolefins.
Charles WEIZMANN
B.P. 572,821, Feb. 2, 1942.

94 Process for the production of acetone and butyl alcohol by fermentation.
Charles WEIZMANN
U.S.P. 2,377,197, Mar. 2, 1942.

95 Improvements in and relating to fermentation processes.
Charles WEIZMANN
B.P. 572,664, Apr. 23, 1941: U.S.P. 2,386,374, May 2, 1942.

96 Improvements in and relating to the preparation and isolation of ethylene chlorhydrin.
Charles WEIZMANN
B.P. 564,090, Aug. 18, 1942; U.S.P. 2,456,350, Nov. 27, 1943 (to Ketarome Ltd.).

97 Improvements in and relating to the aromatisation of hydrocarbonaceous materials.
Charles WEIZMANN
B.P. 575,770, Aug. 18, 1942.

98 Comprehensive aromatisation patent.
Charles WEIZMANN
Trinidad 24, 1942; S. Africa Pat. 456, 1942.

99 Production of higher ketones.
Charles WEIZMANN
S. Africa Pat. 1,028, 1942.

100 Improvements in and relating to the separation of diolefines from gaseous mixtures.
Charles WEIZMANN
B.P. 572,997, Mar. 17, 1943.

101 Improvements in and relating to the production of acetone and butyl alcohol by fermentation processes.
Charles WEIZMANN
B.P. 573,216, Mar. 17, 1943.

102 Improvements in and relating to the separation of butadiene from gaseous mixtures.
Charles WEIZMANN
B.P. 574,058, Mar. 17, 1943.

103 Dehydration of unsaturated alcohols.
Charles WEIZMANN
U.S.P. 2,381,148, Apr. 1, 1943; B.P. 577,364, Mar. 24, 1944.

104 Process for the preparation of alkoxyisobutyric acids.
Charles WEIZMANN
B.P. 578,082, Apr. 15, 1943.

105 Process for the preparation of methacrylic acid and its esters.
Charles WEIZMANN
B.P. 584,607, Apr. 15, 1943.

106 Lining for hydrocarbon treating apparatus.
Charles WEIZMANN and Herbert STEINER
U.S.P. 2,354,163, May 7, 1943.

107 Manufacture of styrene and ethylbenzene.
Charles WEIZMANN
U.S.P. 2,384,984, June 11, 1943.

108 Process for the production of saturated ketones.
Charles WEIZMANN
U.S.P. 2,367,078, July 7, 1943.

109 Process for the removal of acetylenic hydrocarbons from mixtures of gases or vapours.
Charles WEIZMANN
B.P. 580,922, Sept. 1, 1943.

110 Use of ketones for motor fuel.
Charles WEIZMANN
S. Africa Pat. 277, 1943.

111 Use of ketones for motor fuel.
Charles WEIZMANN
S. Africa Pat. 278, 1943.

112 Improved process for carrying out condensation reactions of substances containing reactive methylene or methine groups.
Charles WEIZMANN
B.P. 582,191, Mar. 2, 1944.

113 Production of aromatic hydrocarbons.
Charles WEIZMANN
U.S.P. 2,349,781, May 23, 1944.

114 Production of mono- and disubstituted alkoxyacetic acids and α, β-unsaturated esters therefrom.
Charles WEIZMANN
B.P. 587,545, July 4, 1944.

115 Making acetylenic alcohols and intermediate formed therein.
Charles WEIZMANN (to Polymerisable Products Ltd.)
U.S.P. 2,435,524, Mar. 16, 1945.

116 Aromatization of oil produced in low temperature carbonization of coal.
Charles WEIZMANN (to Ketarome Ltd.)
U.S.P. 2,430,416, May 3, 1945.

117 Process for carrying out condensation reactions.
Charles WEIZMANN (to Polymerisable Products Ltd.)
U.S.P. 2,474,175, May 3, 1945.

118 Improvements in or relating to the manufacture of condensation products from substances containing reactive methylene or methine groups.
Charles WEIZMANN
B.P. 594,182, June 27, 1945.

119 Improvements relating to the production of unsaturated carboxylic acids.
Charles WEIZMANN and Max SULZBACHER
B.P. 596,034, July 16, 1945.

120 Alkoxy isobutyric acid derivatives.
Charles WEIZMANN (to Polymerisable Products Ltd.)
U.S.P. 2,490,109, Jan. 9, 1946.

121 Manufacture of condensation products.
Charles WEIZMANN (to Polymerisable Products Ltd.)
U.S.P. 2,472,135, Aug. 20, 1946.

122 Preparation of carboxylic acids by the oxidation of aldehydes.
Charles WEIZMANN (to Polymerisable Products Ltd.)
U.S.P. 2,456,549, Jan. 29, 1947.

123 Making alkoxy isobutyric acids and derivatives thereof.
Charles WEIZMANN (to Polymerisable Products Ltd.)
U.S.P. 2,525,249, Jan. 12, 1949.

124 Condensation processes.
Charles WEIZMANN (to Polymerisable Products Ltd.)
U.S.P. 2,592,365, Jan. 12, 1949.

Unpublished Passages on Science in the Drafts of Weizmann's Autobiography, *Trial and Error*

Readers familiar with the subject know that Chaim Weizmann decided to suppress a passage in *Trial and Error* that dealt with the crisis in his relations with David Ben-Gurion during World War II and not to publish it at the end of Chapter 41. However, perusal of the folder in the Weizmann Archive containing the manuscript of Weizmann's autobiography reveals that he chose to suppress some other passages as well, or to publish them in an abbreviated form. Four such passages dealing with subjects discussed in this volume appear below. My thanks to Ms. Merav Segal, Director of the Chaim Weizmann Archive, for enabling me to examine this file as well as others in the archive.

The published versions of the passages in question are drawn from the original 1949 edition of *Trial and Error*.[1] In the manuscript versions, words appearing in the published text are in italics. A few egregious spelling errors in the manuscript have been corrected. [Ed.]

I. On Einstein's 1921 journey to the U.S. in the company of Weizmann, and Fritz Haber's objections to it. *Trial and Error*, Chap. 32, pp. 434–435.

Published version:

> It will be remembered that when I made my first visit to America, in 1921, I had been fortunate enough to enlist the co-operation of

1 Chaim Weizmann, *Trial and Error*, London: Hamish Hamilton, 1949.

Einstein. I learned later that Haber had done all he could to dissuade Einstein from joining me; he said, among other things, that Einstein would be doing untold harm to his career and to the name of the Institute of which he was a distinguished member if he threw in his lot with the Zionists, and particularly with such a pronounced Zionist as myself.

Manuscript version, Weizmann Archive:

It will be remembered that when I made my first visit to America, in 1921, I had been fortunate enough to enlist the co-operation of Einstein. It was something of a triumph for us, and ensured in advance that the Hebrew University in the making – which Einstein was to take as his field on the visit – would receive the right kind of attention. Everything was set for the trip, and we were to meet Einstein in Plymouth, which he was to reach on a Dutch boat; thence we were to proceed together. A few days before Einstein's departure we learned from some friends in Berlin that he was subjected to great pressure by certain of his colleagues in Dahlem, where he lived, to cancel his trip. Chief among these men was Haber, who used all his energy and eloquence to persuade Einstein that he *would be doing untold harm to his career and to the name of the Institute of which he was a distinguished member if he threw in his lot with the Zionists, and particularly with such a pronounced Zionist as myself.* I hardly knew Einstein at the time, and was genuinely disturbed until I learned that this sort of argument made no impression on him.

II. From the end of Weizmann's description of his visit with Fritz Haber in Dahlem, Berlin, at the end of 1932. *Trial and Error*, Chap. 32, p. 436.

Manuscript version, Weizmann Archive:

During the lunch at Haber's home I turned to him and said: "Your Institute is a great one, marvelous in its program, its equipment and its staff. But it suffers in my humble opinion from one fundamental defect – there are too many Jews in it. Whatever section I came into, the head was a Jew, his first and second and third assistants were Jews, and somewhere at the bottom of the ladder were a few non-Jews. Imagine," I went on, "an institute like that in Palestine, with an Arab at the head, all the important posts occupied by Arabs, and the very subordinate positions allotted to Jews. It's unthinkable. Yet to all intents and purposes you have this abnormal structure here. I don't want to be a Cassandra, but I can't help predicting that one day the subordinates will rise and throw their Jewish chiefs out."

Haber looked at me in astonishment and perplexity, then said: "Dr. Weizmann! You are an anti-Semite! You mustn't say such things. My position in this institute is unchallengeable."

There passed through my mind the story of Willstätter and the university in Munich, but I did not want to enter into an argument with my host.

I left Dahlem heavy-hearted and filled with forebodings, which I remember communicating to my wife on my return to London.

III. On the accomplishments of Jewish scientists and the failure to recognize them as such, which would be rectified by the development of science in Palestine. *Trial and Error*, Chap. 32, p. 438, following Weizmann's description of his efforts to persuade Richard Willstätter and Fritz Haber to emigrate to Palestine.

Published version:

These were two of the men whom I sought to attract to our institutions in Palestine, both for their sake and for ours. There were others, of course. I felt it would be a great accession of moral strength, and a valuable source of technical knowledge if we could offer to the Hebrew University, or to the Sieff Institute, Albert Einstein the physicist, James Franck of Göttingen, the mathematician Hermann Weyl, the physicist Placzek, the chemist Wiegener, to mention but a few names. But somehow I failed to convince them.

Manuscript version, Weizmann Archive:

These were two of the men – I shall tell of others – *whom I* wanted *in Palestine*. But they preferred to serve other gods.

Very few people – Jewish or non-Jewish – have any notion of the role which Jews have played in modern science, and particularly of their astounding share in the development of nuclear physics. I would not like to venture a numerical estimate, but I have heard Einstein speak of ninety percent. In any case, reading on the literature of this and kindred subjects, I am continuously struck by the utter disproportion of the Jewish contribution. For instance, H.D. Smyth, in his "Atomic Energy for Military Purposes" (*Modern Physics*, Vols. 16–17, 1944–45, p. 373), states: "At that time (January 1939) American-born nuclear physicists were so unaccustomed to the idea of using their science for military purposes that they hardly realised what was needed to be done. Consequently the early efforts at restricting publication and at getting government support were stimulated largely by a small group of foreign-born physicists centering on L. Szilard, and including E. Wigner, E. Teller, V.F. Weisskopf and E. Fermi." All of these, with the exception of the last-named, are Jews.

It would of course be absurd for the fact to be mentioned in a scientific report. But it is seldom mentioned anywhere else either. I must always think of Einstein's words to me during our trip to America. He said: "If my theory of relativity turns out to be correct, it will be put down as a great achievement of German scientists. If it proves to be wrong it will be the damned Jews who misled the world of science." If human folly reaches such a stage that atomic energy will be used extensively in the next war about which one hears so much talk, it will be said that the Jews have conspired to destroy the world. If, however, as I hope and believe will be the case, atomic energy will be guided into constructive channels, and humanity will enjoy the benefits of unlimited sources of energy, solving all its problems of fuel and transportation, I doubt whether people will remember the great number of Jews who will have helped to bring these results about. With the exception of a few anti-Semites no one thought of Lilienthal, who became the head of what is probably the greatest atom research station in the world, as a Jew.

Without wishing to see in everything a justification of my Zionist credo, I cannot help reflecting on such matters and arriving at the conclusion that each nation keeps its own account in the general ledger of humanity, and enters its achievements into the record of civilisation. We Jews alone do not keep our own account; it is kept for us by others; therefore the entries figure under different names, and our contributions become invisible. It is only in Palestine, where we are building up our normal life, that we begin to keep our own accounts; and this is why Jewry instinctively feels that what we do or fail to do here is likely to reflect credit or discredit on Jews in America or Australia or elsewhere, who may not be taking a direct part in the actual performance.

But, as I have intimated, a great many of the scientists – among others – did not see things my way; I was anxious to secure their services for Palestine, *for their sake and ours. I felt* that *it would be a great accession of moral strength, and a valuable source of technical knowledge, if we could offer to the Hebrew University, or to the Sieff Institute, Albert Einstein, the physicist James Franck of Goettingen, the mathematician Hermann Weyl, the physicist Placzek, the chemist Wiegener,* and others of their standing. *But somehow I failed to convince them.*

IV. Weizmann's research program, a shortened version of which appears in *Trial and Error*, Chap. 43 (pp. 545–547). In the printed edition, the original title, "Zionism and Science," was changed to "Science and Zionism."

Manuscript version, Weizmann Archive. Words that appear in the published version but not in the manuscript are in square brackets.

I referred, in the last chapter, to a friendly meeting with Mr. Ernest Bevin – one in which he sought my services. It occurred in the midst of the war, when the British Government sent out to West Africa a small commission to investigate the short- and long-range possibilities of new sources of raw materials[*, with fuel chiefly in view*]. *Walter Elliot and Creech-Jones were on the commission, and I had several subsequent meetings with them.*

I suggested that they try to determine whether large quantities of *various types of starches could not be grown* [*easily*] *in West Africa. It is known that Central or tropical Africa produces a great many root starches, like manioc and tapioca; also cane sugar. I was of the opinion that if one could grow abundant supplies of these commodities, one could introduce a fermentation industry into that part of the world, with a large yield of ordinary alcohol, both for power and for the production of butyl alcohol and acetone. These three materials, in large quantities and at a low price, could form the basis of two or three great industries, among them high-octane fuel, and would make the British Empire independent of oil wells.*

The commission went out for a survey, and so far nothing has come of it. But *I am still of the opinion that the plan is feasible.* Nor are its advantages confined to fuel. The materials mentioned would form the basic starting points for a great plastics industry. I have mentioned that in regard to fuel, the British Empire could become independent of oil wells, which are in fixed localities. The [*Its*] *most attractive feature* [*is*], *perhaps,* of a good fermentation industry, consists in the fact *that it is not tied to a geographic point,* [*like an oil supply, but is applicable wherever the substances I have mentioned can be grown*]. So, for instance, one could have a great fermentation plant in Simonstown, near the Cape, which is central to the Atlantic and Indian oceans, and it would not be necessary to transport the fuel for the navy all the way from Arabia. Fermentation plants could also be set up at points in India, where the raw material would be rice. At the beginning only a certain proportion of the demand could be met in this way, but once started there is no reason why it should not be developed till it meets one hundred per cent of the demand, thus breaking the world's oil monopoly. From the chemical point of view, the plan is in my opinion possible and realisable; it would undoubtedly meet with the opposition of vested interests, but this could be overcome by a government where private enterprise might fail.

Much of my work and that of my colleagues has been directed to [*It is, moreover, part of*] *what I believe is a necessary and* [*probably*] perhaps *inevitable shift in a* [*great*] large *sector of modern industry.* The three substances I have mentioned, *butyl alcohol, acetone, and ethyl alcohol are the bases of* [*many*] a large variety of *products* [*besides fuel and plastics. The acetylene chemistry derivatives start with methyl butinol, which is itself prepared from acetylene and acetone. Methyl alcohol is made from carbon dioxide and hydrogen, which are yielded as by-products from the fermentation of butyl alcohol; the*] Butyl alcohol fermentation, for instance, yields, as a by-product, large quantities of carbon mon-oxide and hydrogen, out of which one could make methyl alcohol, or if necessary, methane, which are themselves starting materials for other valuable products. *Methyl alcohol can be easily reconverted into formaldehyde, one of the best disinfectants. It is, moreover, widely used in synthetic chemistry.*

In other words, the fermentation industry could yield a range of substances, all made out of starch, leading to plastics, fuel and various chemicals. A whole set of substances can be made of methyl butinol, which is prepared easily and in good yield from acetylene and acetone. Acetylene can be easily obtained by the ordinary method from calcium carbide, or else from methane. The acetylene chemistry derivatives, too, start with methyl butinol; the process is being worked at present on a pilot plant scale, in America, in my laboratory in London, and in the Sieff Institute, in Rehovot.

Methyl butinol, again, leads, by a simple reaction, *to the formation of certain amino-alcohols, which are most valuable constituent of dyestuffs.* By combining these alcohols with certain derivatives of anthra-quinone, one can obtain red and blue dyes which, I am informed, are so far the best and fastest dyes known. [*It is on these lines that my collaborators and I have been working for a number of years. The programme*] All this work *is still in its initial stages, and to elaborate* [*it*] the program *would require quite a number of chemists, and a certain* [*amount*] period *of time, but enough has been done – in Philadelphia,* in *London and* at *Rehovot – to indicate the lines of research on which we should move at present.*

One of these lines, [*Another piece of work*] *which has* [*been occupying*] occupied [*our*] my *attention* and that of my collaborators [*for*] in *the last few years leads to the production of cheap but digestible and valuable nutritive products,* made from the residues of certain industries. *It may be briefly described as the attempt to upgrade materials which are used as cattle food, converting them into human food. The materials are peanuts or peanut cake – after the oil has been extracted – soya beans and similar* [*substances*] products. These, when treated enzymatically, are solubilised and degraded into low molecular proteins. The enzymes are obtained from yeast, so if yeast and peanut butter are treated together in a certain way one obtains a palatable food rich in vitamins. The whole process is carried out at a low temperature, not higher than fifty degrees, so that all the valuable ingredients of the yeast remain intact. *This product has been tested in many hospitals as nutrition for patients, and for people with ulcerated stomachs, and it has proved very beneficial. It is, moreover, cheaply produced, and is within the reach of the poorest populations, such as the coolies of* [*India or China*] China or India. A little pilot plant was, in fact, set up in India by a firm with which I was connected, but the whole production has now been transferred to an American firm. [*It is entirely of a vegetable nature, is highly nutritive, and without containing a particle of meat has a meaty taste. It*] The product *should be of particular benefit to those eastern countries in which meat is either too expensive or is prohibited for religious reasons.* It is entirely of a vegetable nature, is highly nutritive, and without containing a particle of meat has a meaty taste. If an Indian coolie could supplement his regular diet with a little tube of this stuff – which should sell at a fraction of a cent – it would make a tremendous difference to him.

[*This enterprise was worked out*] All this side of research is being studied *in its technical aspect by a group of capable workers in America, and* [*has already produced results*] should come to fruition very soon. *My colleagues and I have been occupied with the chemical side since 1935, aided at the beginning by Willstätter, who was a great authority on the chemistry of proteins.*

A third branch of research has occupied my attention in recent years, and in this Dr. David Bergmann and one or two others have participated to a considerable degree. In its early stages it was carried out at the Sieff Institute, later it was transferred to London. It is a process for converting the crude residues obtained after the distillation of oil – the high-boiling constituents of petrol after the low-boiling gasoline has been removed. By a catalytic process of great simplicity and considerable efficacy the crude material is converted *into aromatic substances like benzine, toluene, xylene, naphthalene and certain gaseous substances like butylene, isobutylene, and so on.* One hundred parts of the crude oil, if cracked by means of the catalyst, yields fifty parts of gaseous yield as above. Both yields are of great value.

The process was worked out during the war, first, as I have said, at the Sieff Institute in Rehovot, and then at my laboratory in London. *Our purpose was to create reserves of toluene, for we remembered our sad experiences in the last war, when we ran out of toluene, the basis of T.N.T. The process proved*[*, however,*] *to have* a *wider value* because of the other constituent parts, which are obtained in a very pure state, not needing a complicated process of rectification, as is the case when they are obtained from coal tar. Moreover, the yields are larger, and the whole thing is simpler. The process was examined and re-examined during the war, and found of great interest. *It was taken up by a private firm to which the Manchester Oil Refinery belongs; a company was formed with a capital of two million pounds*[*, the government participating to the extent of 50 per cent*] with fifty per cent government participation *in view of the national importance of the process.* A great plant is at present in construction in Trafford Park, Manchester, and should be in operation by April, 1948.

These are my main practical interests, apart from some purely scientific researches.

It has been a long struggle, partly against inertia, and partly against vested interests, to put the enterprise on a proper footing, but I hope it is now over. I have made up my mind not to engage in any more researches which may have technical importance. For that one has to be much younger, and to possess the necessary resilience in order to fight one's way through the maze of difficulties. The one piece of work which still requires completion is the fermentation of cellulose, straw, wood, sawdust, wood residues, etc.

It is well known that cellulose can be saccharified and thus fermented, but a clean and easy method of saccharification, technically simple and accessible, is still to be worked out. Once this problem is solved, it will be possible to produce alcohol, acetone and all the other above-mentioned substances from wood instead of from starch. The advantage is obvious. There are millions and millions of tons of cellulose in the form of wood, leaves, branches, grass, straw, etc. going to waste every year. If this can be worked up properly then all the derivatives would be obtainable at a low price. It is still a practical problem, and a great many people are trying to solve it.

The ideas set forth in brief outline in the foregoing pages were germinating in my mind for many years.

V. In one of the drafts of the book preserved in the Weizmann Archive, Weizmann crossed out several sentences relating to his scientific work in Switzerland and reformulated them, correcting several significant details and adding some additional ones. This revised version was never published; the original version would ultimately appear, unaltered, in *Trial and Error*, pp. 76–77.

Published version:

My doctorate thesis was based on the dyestuff researches I had started in Berlin, and on the discovery which I had tried unsuccessfully to sell in Moscow. I managed to obtain with my doctorate the coveted top rating of *summa cum laude*, and the autumn following my graduation I was appointed *Privat Dozent* in chemistry at the University of Geneva. The nearest equivalent to this post in an English university is that of assistant lecturer. There is one important difference. The *Privat Dozent* received no fixed salary. He was paid by the pupil, at about fifty marks per term. The average enrolment gave the *Privat Dozent* something less than a very modest livelihood. But of course the title carried with it a certain distinction. It was the beginning of an academic career. It afforded opportunity for study and research. The next step was an assistant professorship, and after that came a full professorship. So, with all its poor pay, the post of *Privat Dozent* was much coveted.

I do not know how long I would have been able to retain the lectureship if it had not been for a great stroke of luck almost at the beginning. I was able to sell a patent to the *I. G. Farbenindustrie* of Germany, and this provided me at once with a regular income of six hundred marks a month. That was a tremendous experience for me. I had become independent! What was more, I had achieved independence by my own efforts, and in my own field as chemist.

As to the actual contract with that gigantic enterprise, the *I. G. Farbenindustrie*, I paid little attention to it. Hardly anyone thought of it then as the focus of German military might and of German dreams of world conquest. But it gives me a queer feeling to remember that I too, like many another innocent foreign chemist, contributed my little to the power of that sinister instrument of German ambition. [...]

Manuscript version, Weizmann Archive:

My doctorate thesis was based on some of the researches I started in Berlin. I managed to pass my degree with high marks[2] and the autumn following my graduation I have decided to settle in Geneva which was famous for its chemical school. Here I began research into a subject which in later years has occupied me a good deal. After having published the first results I could apply for the post of

2 Weizmann deleted the words: "to obtain with my doctorate the coveted top rating of *summa cum laude*," which ultimately appeared in the book.

Privat-Dozent. The nearest equivalent to this post in an English or American university is, I think, that of lecturer. In Geneva it meant, at least then, much less than in a German or English university. The Privat-Dozent [received] no salary. The only emolument derived from the fees which the students paid who happened to take this particular course and they were very few as a rule. [...]

I do not know how long I would have been able to remain in Geneva to which I was greatly attached if not for a great stroke of luck almost at the outset. My research yielded a practical result in which the Farbenfabriken Bayer in Elberfeld were interested and this provided me at once with a regular income of about four[3] hundred marks a month. [...]

As to the actual contract with that gigantic enterprise, the I.G. Farbenindustrie, I paid little attention to it. Hardly anyone thought of it then as the focus of German military might and of German dreams of world conquest. Now it gives me a somewhat queer feeling to remember that I too, like many another innocent Jewish[4] chemist, had some relation with that instrument of German ambition. It is still more strange and in a sense very tragic to think that the scientific achievements of Jewish chemists in Germany have contributed greatly to the efflorescence of the great chemical industry in that country. Names like Haber, Willstätter, Ladenburg, Wallach, Caro, Ehrlich and his collaborators represent the more prominent in a whole gallery of scientists.

3 Weizmann deleted the word "six," which would appear in the book.

4 Weizmann crossed out the word "foreign" and replaced it with "Jewish."

ILLUSTRATIONS

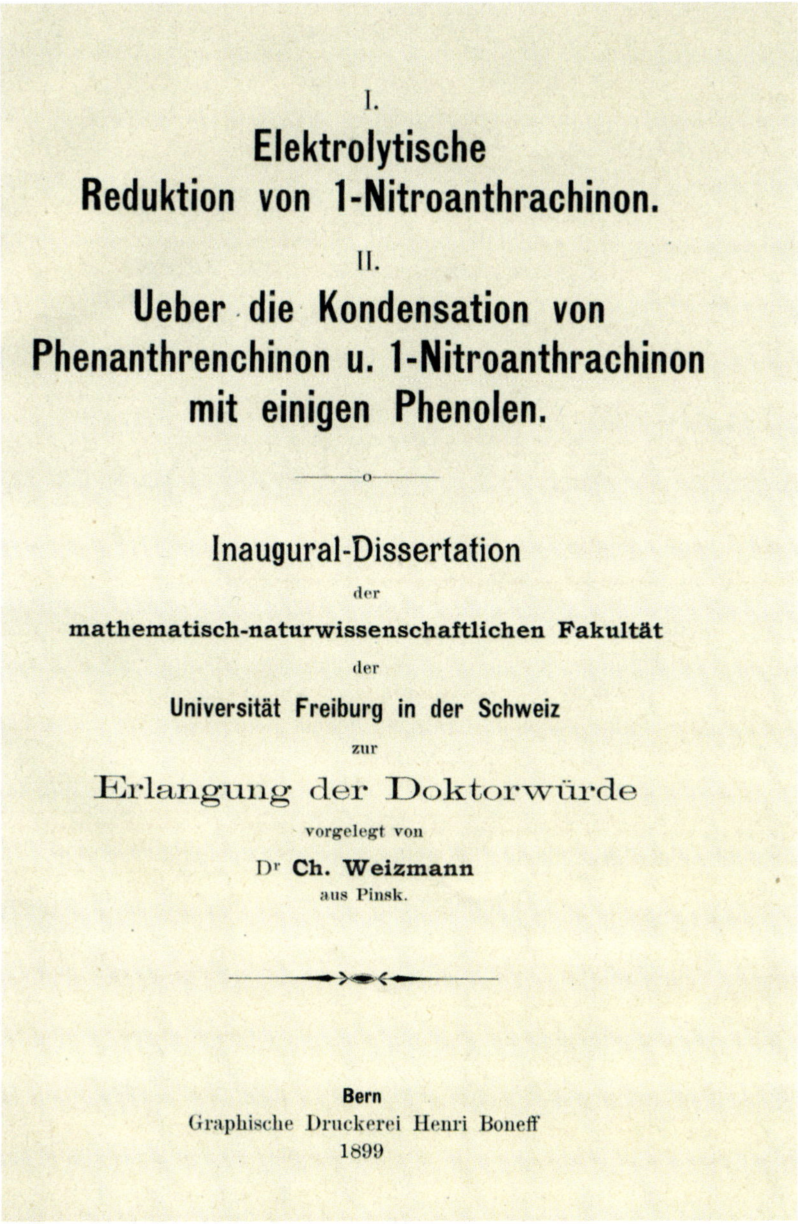

I.

Elektrolytische Reduktion von 1-Nitroanthrachinon.

II.

Ueber die Kondensation von Phenanthrenchinon u. 1-Nitroanthrachinon mit einigen Phenolen.

Inaugural-Dissertation

der

mathematisch-naturwissenschaftlichen Fakultät

der

Universität Freiburg in der Schweiz

zur

Erlangung der Doktorwürde

vorgelegt von

Dr **Ch. Weizmann**

aus Pinsk.

Bern

Graphische Druckerei Henri Boneff

1899

1. Title page of Weizmann's Ph.D. dissertation, University of Freiburg, Switzerland, 1899 (courtesy of the CWA)

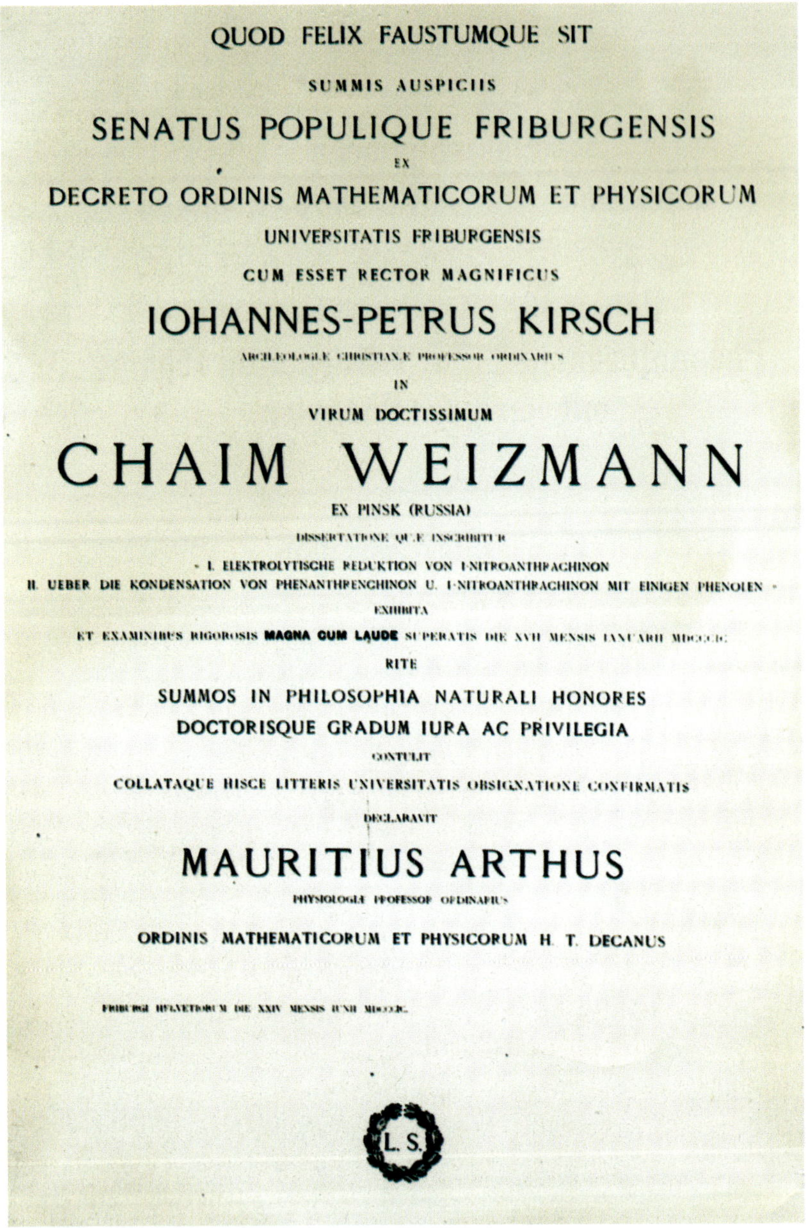

QUOD FELIX FAUSTUMQUE SIT

SUMMIS AUSPICIIS

SENATUS POPULIQUE FRIBURGENSIS

EX

DECRETO ORDINIS MATHEMATICORUM ET PHYSICORUM

UNIVERSITATIS FRIBURGENSIS

CUM ESSET RECTOR MAGNIFICUS

IOHANNES-PETRUS KIRSCH

ARCHAEOLOGIAE CHRISTIANAE PROFESSOR ORDINARIUS

IN

VIRUM DOCTISSIMUM

CHAIM WEIZMANN

EX PINSK (RUSSIA)

DISSERTATIONE QUAE INSCRIBITUR

» I. ELEKTROLYTISCHE REDUKTION VON 1-NITROANTHRACHINON
II. UEBER DIE KONDENSATION VON PHENANTHRENCHINON U. 1-NITROANTHRACHINON MIT EINIGEN PHENOLEN »

EXHIBITA

ET EXAMINIBUS RIGOROSIS **MAGNA CUM LAUDE** SUPERATIS DIE XVII MENSIS IANUARII MDCCCIC

RITE

SUMMOS IN PHILOSOPHIA NATURALI HONORES

DOCTORISQUE GRADUM IURA AC PRIVILEGIA

CONTULIT

COLLATAQUE HISCE LITTERIS UNIVERSITATIS OBSIGNATIONE CONFIRMATIS

DECLARAVIT

MAURITIUS ARTHUS

PHYSIOLOGIAE PROFESSOR ORDINARIUS

ORDINIS MATHEMATICORUM ET PHYSICORUM H. T. DECANUS

FRIBURGI HELVETIORUM DIE XXIV MENSIS IUNII MDCCCIC

L. S.

2. Weizmann's Ph.D. diploma, University of Freiburg, Switzerland, 1899 (courtesy of the CWA)

3. Weizmann in his laboratory in Geneva, 1902
(courtesy of the CWA)

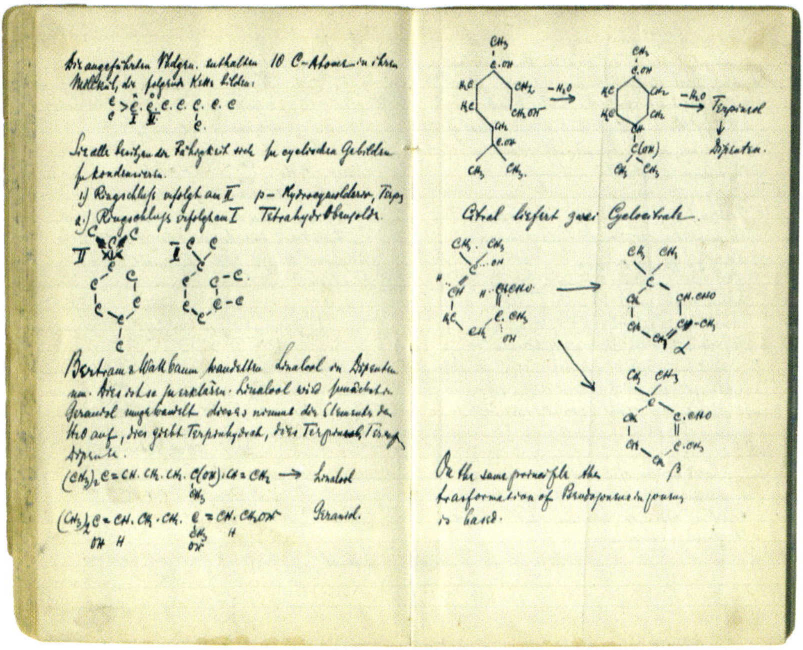

4. Two pages in Weizmann's chemistry laboratory notebook, University of Manchester, 1906
(courtesy of the CWA)

5. Weizmann in his laboratory at the University of Manchester, 1912
(courtesy of the CWA)

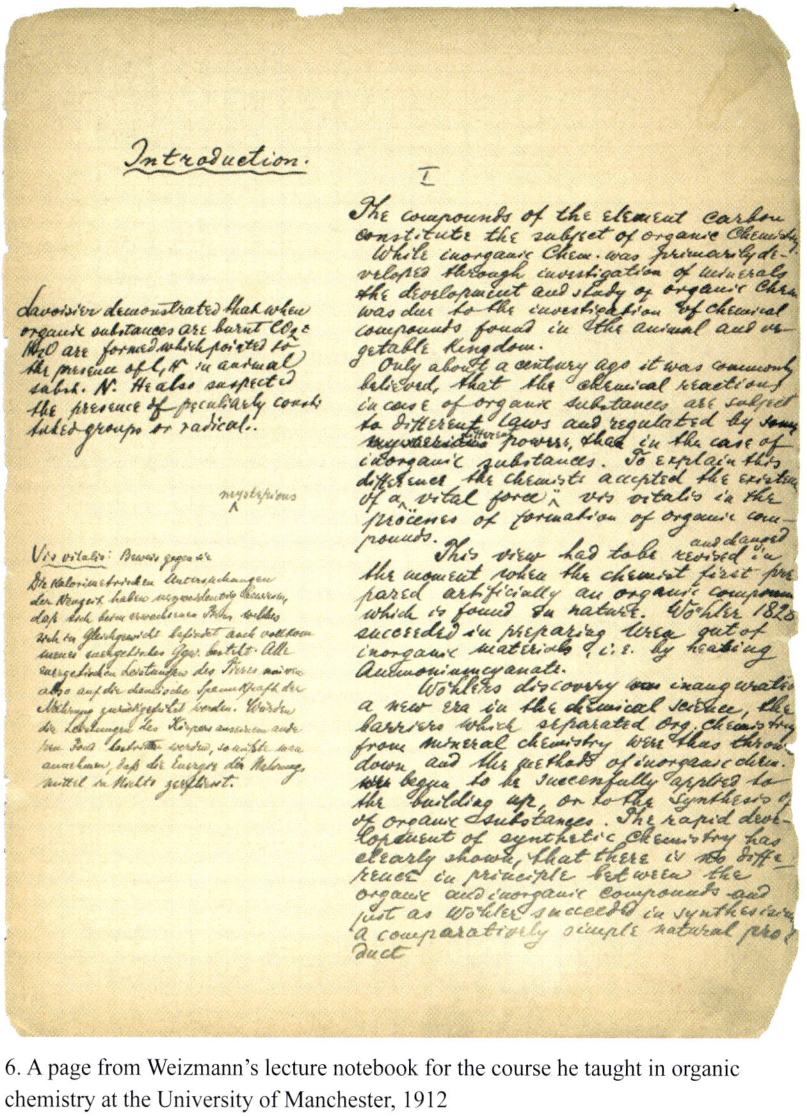

Introduction.

I

The compounds of the element carbon constitute the subject of organic Chemistry. While inorganic Chem. was primarily developed through investigation of minerals the development and study of organic Chem was due to the investigation of chemical compounds found in the animal and vegetable kingdom.

Only about a century ago it was commonly believed, that the chemical reactions in case of organic substances are subject to different laws and regulated by some mysterious powers, than in the case of inorganic substances. To explain this difference the chemists accepted the existence of a vital force ^ vis vitalis in the processes of formation of organic compounds.

This view had to be revised and changed the moment when the chemist first prepared artificially an organic compound which is found in nature. Wöhler 1828 succeeded in preparing urea out of inorganic materials i.e. by heating Ammoniumcyanate.

Wöhlers discovery was inaugurating a new era in the chemical science, the barriers which separated org. chemistry from mineral chemistry were thus thrown down, and the methods of inorganic chem. begun to be succesfully applied to the building up, or to the synthesis of organic substances. The rapid development of synthetic Chemistry has clearly shown, that there is no difference in principle between the organic and inorganic compounds and just as Wöhler succeeded in synthesising a comparatively simple natural product

Lavoisier demonstrated that when organic substances are burnt CO_2 & H_2O are formed which pointed to the presence of C, H in animal subst. N. He also suspected the presence of peculiarly constituted groups or radicals.

mysterious ^

Vis vitalis: Beweis gegen sie
Die kalorimetrischen Untersuchungen der Neuzeit haben unzweideutig bewiesen, daß sich beim erwachsenen [illegible], welches sich im Gleichgewicht befindet auch vollkommen [illegible] Gew. [illegible]. Alle [illegible] Leistungen des Tieres [illegible] also auf die chemische Spannkraft der Nahrung zurückgeführt werden. Würden die Leistungen des Körpers [illegible], so müßte man annehmen, daß die Energie der Nahrungsmittel in Nichts zerfließt.

6. A page from Weizmann's lecture notebook for the course he taught in organic chemistry at the University of Manchester, 1912
(courtesy of the CWA)

7. The ceremony for laying the foundation stones of the Hebrew University of Jerusalem, 15 Av 5678/July 24, 1918
(courtesy of the Hebrew University)

8. The foundation stones of the Hebrew University, 15 Av 5678/July 24, 1918
(courtesy of the Hebrew University)

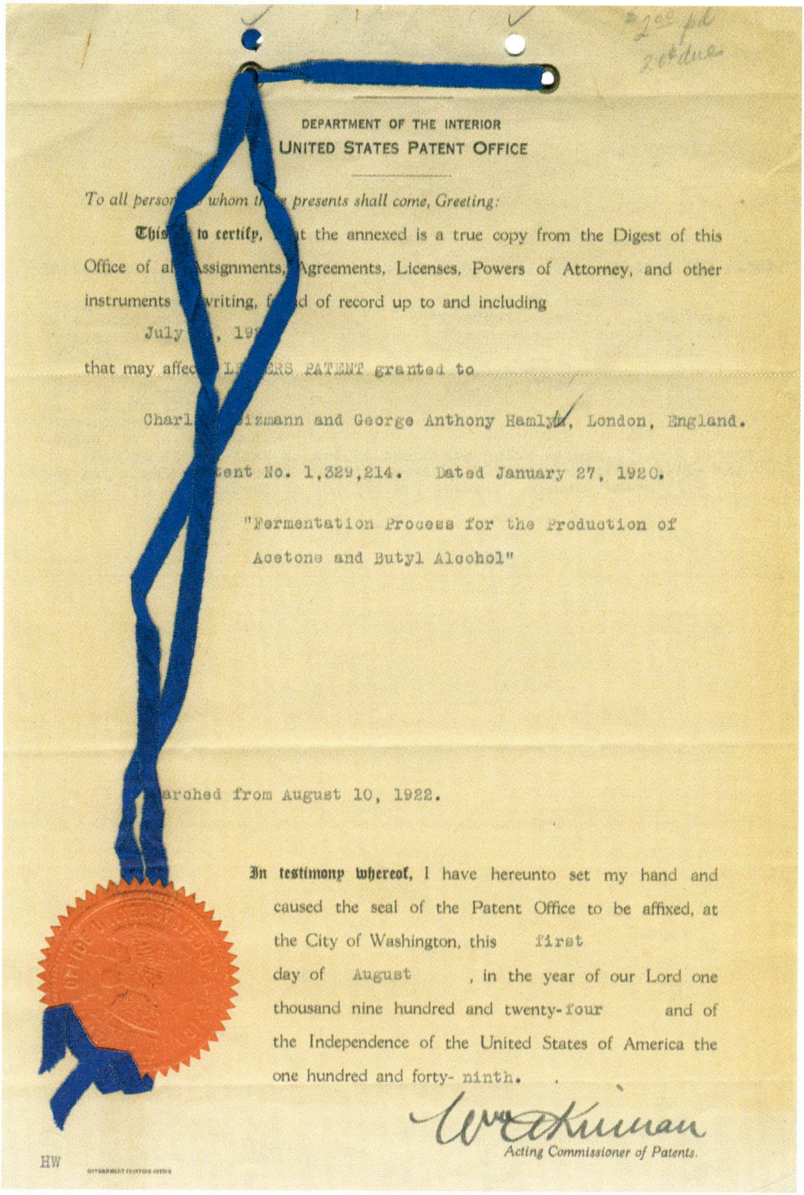

$2.00 pd
20¢ due

DEPARTMENT OF THE INTERIOR
UNITED STATES PATENT OFFICE

To all perso[…] whom th[…] presents shall come, Greeting:

This […] to certify, […]t the annexed is a true copy from the Digest of this Office of a[…] Assignments, Agreements, Licenses, Powers of Attorney, and other instruments […]writing, f[…]d of record up to and including

July […], 19[…]

that may affec[…] L[…]ERS PATENT granted to

Charl[…] […]izmann and George Anthony Hamlyn, London, England.

[…]ent No. 1,329,214. Dated January 27, 1920.

"Fermentation Process for the Production of Acetone and Butyl Alcohol"

[…]arched from August 10, 1922.

In testimony whereof, I have hereunto set my hand and caused the seal of the Patent Office to be affixed, at the City of Washington, this first day of August, in the year of our Lord one thousand nine hundred and twenty-four and of the Independence of the United States of America the one hundred and forty- ninth.

Wm. A. Kinnan
Acting Commissioner of Patents.

HW

9. Patent certificate for Weizmann's fermentation process for the production of acetone and butyl alcohol, January 27, 1920
(courtesy of the CWA)

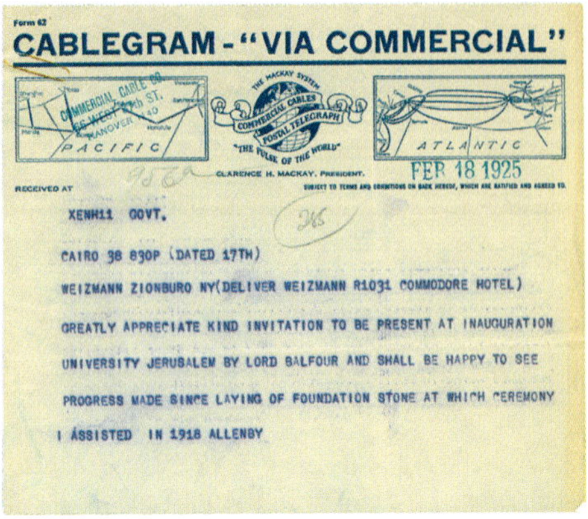

Form 62

CABLEGRAM - "VIA COMMERCIAL"

FEB 18 1925

RECEIVED AT

XENH11 GOVT.

CAIRO 38 830P (DATED 17TH)

WEIZMANN ZIONBURO NY(DELIVER WEIZMANN R1031 COMMODORE HOTEL)

GREATLY APPRECIATE KIND INVITATION TO BE PRESENT AT INAUGURATION UNIVERSITY JERUSALEM BY LORD BALFOUR AND SHALL BE HAPPY TO SEE PROGRESS MADE SINCE LAYING OF FOUNDATION STONE AT WHICH CEREMONY I ASSISTED IN 1918 ALLENBY

10. Cablegram sent by Allenby to Weizmann on February 17, 1925, thanking him for his invitation to the Hebrew University's opening ceremony
(courtesy of the CWA)

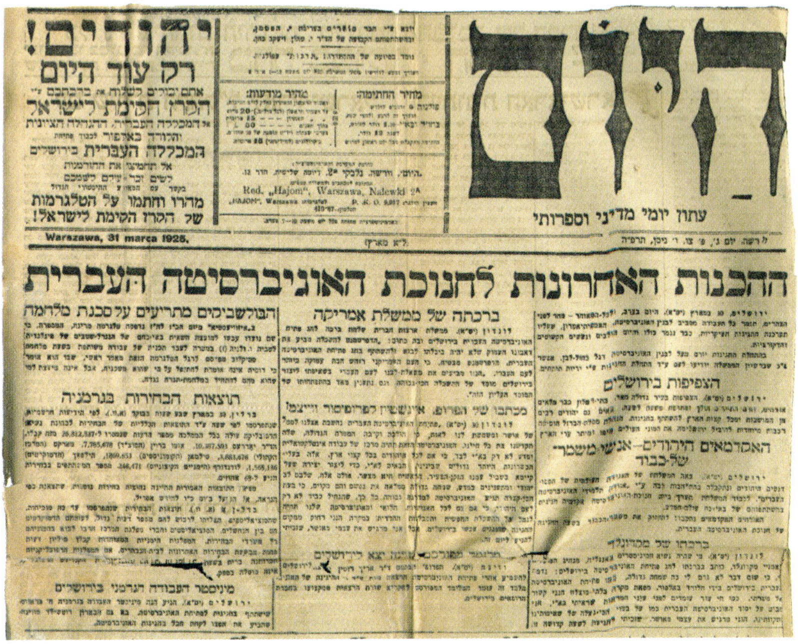

היום

עתון יומי מדיני וספרותי

יהודים! רק עוד היום

Warszawa, 31 marca 1925.

ההכנות האחרונות לחנוכת האוניברסיטה העברית

ברכתה של ממשלת אמריקה

הבולשביקים מתריעים על סכנת מלחמה

תוצאות הבחירות בגרמניה

הצפיפות בירושלים

האדמו"רים היהודים – אנשי משמר של־כבוד

ברכתו של מקדונלד

מיניסטר העבודה הגרמני בירושלים

11. Article on the preparations for the dedication of the Hebrew University of Jerusalem, *Hayom*, March 31, 1925
(courtesy of the CWA)

העולם

עתונה המרכזי של ההסתדרות הציונית העולמית

יוצא אחת לשבוע בהשתתפותו הקבועה של נחום סוקולוב.
העורך: משה קלינמן.

גליון י״ד-ט״ו ∴ לונדון, ט׳ ניסן, תרפ״ה ∴ LONDON, APRIL 3rd, 1925 ∴ שנה שלש־עשרה

לפתיחת המכללה העברית בירושלים

חיים וויצמן

תרע״א - תרפ״ה

את הרעיון הראשון על האוניברסיטה העברית הגה, כנודע, לראשונה הפרופיסור הרמן שפירא ז״ל, ואנו נותנים בגליון זה את תמונתו ומאמר מענין עליו מאת הד״ר י. ל. ברוך. בזמן ההוא לא היתה עוד ההסתדרות הציונית מוכשרה לתפוס את הדבר בענין ממשי שאפשר לגשת להוצאתו אל הפועל, או אפילו רק להתחיל להתבונן אליו, והרעיון נשאר תלוי באויר.

לא עברו ימים רבים והרעיון מצא לו גואל חדש—את הד״ר חיים וויצמן. את ערכו הנעלה של רעיון האוניברסיטה הכיר לבסוף גם הרצל בכבודו ובעצמו וזה אשר כתב במכתב מיום 20 יולי 1902:

„ענין האוניברסיטה מוסיף לתפוס את לבי במדה רבה. אל יאין בי ארוני לחזור ולומר זאת. יאיר נא לי אדוני להגיד לו בשיחה בע״פ בשעת פגישתנו הקרובה, איך אני מצייר לי את הדבר״.

מועקת התלמידים היהודים בבתי־המדרשות של אירופה, שהשתלטה בהם האנטישמיות בכל תקפה, וביחוד ענוייהם של צעירי היהודים ברוסיה, שבתי־המדרש הגבוהים במעט נסגרו בפניהם כליל, הוסיפו לויה של „צורך השעה״ לפרובלימה ע״ד יסוד אוניברסיטה עברית לאומית. וויצמן העמיס על שכמו את הסבל להכין לצורך זה את הלבבות ולהכשיר את הקרקע להגשמתו. עוד רב היה העמל לרכוש את תשומת־הלב הראויה לרעיון זה אפילו בקרב הציונים. מקצתם היו קטני־אמנה וחשבו את כל הענין ל„חלום פורח״. מקצתם העו בדבר אם ימצאו לנו האמצעים והכחות הדרושים. ורבים, ביחוד מתוך „הציונים המעשיים״, היו אשר גם התנגדו בגלוי וטענו, שחטא נחטא לישוב אם נוציא כחות ואמצעים בשביל יסוד בית־מדרש גבוה ל„תועלת פרחי המשכילים״ תחת אשר עלינו לאזור את כל הכחות בשביל הרחבת והתפתחות ההקלאות שלנו בארץ — טענה שעוד היום יש לה מהלכים באילו חוגים של צבורנו...

בכל זאת הולך הרעיון וכובש את הלבבות, ובמדה רבה הודות לעמלו ויגיעתו הבלתי פוסקת של וויצמן, עד שבישיבת הועד הפועל הגדול אשר לפני הקונגרס הי״א, בחודש מרס של שנת תרע״א, החלט להעמיד את הדבר על הפרק בסדר היום של הקונגרס הבא. הקונגרס התקיים בספטמבר של שנת תרע״א בווינה וההרצאה על דבר האוניברסיטה ניתנה לידי הד״ר וויצמן. אנו חושבים לדבר בעתו לתת פה אילה קטעים מן ההרצאה הזאת.

„...כולנו מרגישים מה גדול היה ערכו של מרכז אינטלקטואלי,

[illegible]

כל מעצור ע״י הגבלות, בתוך סביבה יהודית, חפשית מלחצן של תרבויות זרות, בשאיפה כבירה ליצור ערכין חדשים ולקשר את מסורותינו הגדולות עם החדש ביותר. מתוך סינתיזה זו היתה צומחת השכלה יהודית אמתית, שהיתה ראשית־כל לברכה לאומה היהודית. ממרכז כזה היתה הגולה יונקת השפעה רבה, ועל ידו היתה ההכרה־העצמית של אנשי־הרוח בקרבנו מתעלה לאין שעור. אם נשים אל לב, מה גדולה היא כבר גם ההשפעה של הגימנסיון העברי על הגולה, ומה עצום הוא הרושם של ההסתדרות הצעירה והרופסת־עדיין „מרכז המורים״ — אז נבין את גודל הערך הלאומי של אוניברסיטה עברית.

חיים וויצמן

אבל קודם כל יעשה בית־המדרש הגבוה לנוטרתם של אותם נכסי האומה, היקרים לה ביותר בשביל עתידה: הוא יהיה למקום גדולה וטפוחה של הלשון הלאומית שלנו, נקודת־האחוד של כל היצירה היהודית בשדה הספרות, האמנות והמדע, במלה אחת — המרכז התרבותי.

איזה עושר של כח־יצירה יהודי היה יכול להתפתח, לחיות את חייו, בבית־המדרש הגבוה היהודי! — — — רכוז זה של העבודה היהודית המדעית אין ערוך לתועלתו בעד המלומדים עצמם ובעד המדע הכללי.

רעיון החרות העצמית, האובד כ״כ הרבה מהאלמנטים הצעירים שלנו, מחסרון התרבזות ומתוך ההשפעה המזיקה של הסביבה הזרה, היה מוצא את המשמר הכי חזק באותו הנוער, שהיה מפתח באופן הכי נעלה את הרוח הלאומי והיה נושא אותו אל תוך העם. אז היה נוצר לנו היהודי החדש, היוצר, זקוף־הקומה, והעם היהודי היה שואב

12. Article on the opening of the Hebrew University of Jerusalem, *Ha'olam*, April 3, 1925
(courtesy of the CWA)

13. Weizmann at the tenth anniversary celebration of the Hebrew University of Jerusalem, 1935
(courtesy of the Hebrew University)

14. Weizmann's laboratory at the Sieff Institute
(courtesy of the CWA; photograph © Photo Schleisner)

15. Weizmann with the Sieff Institute staff during a visit by Carl Neuberg, 1935
(courtesy of the CWA; photograph © Photo Schleisner)

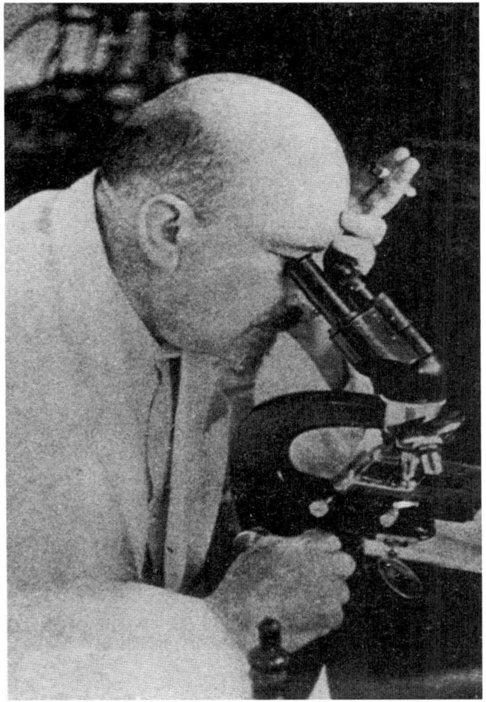

16. Weizmann in his laboratory at the Sieff Institute, 1938
(courtesy of the CWA)

17. Weizmann in his laboratory at the Sieff Institute, 1938
(courtesy of the CWA)

18. The Sieff Institute building
(courtesy of the CWA; photograph © the Weizmann Institute of Science)

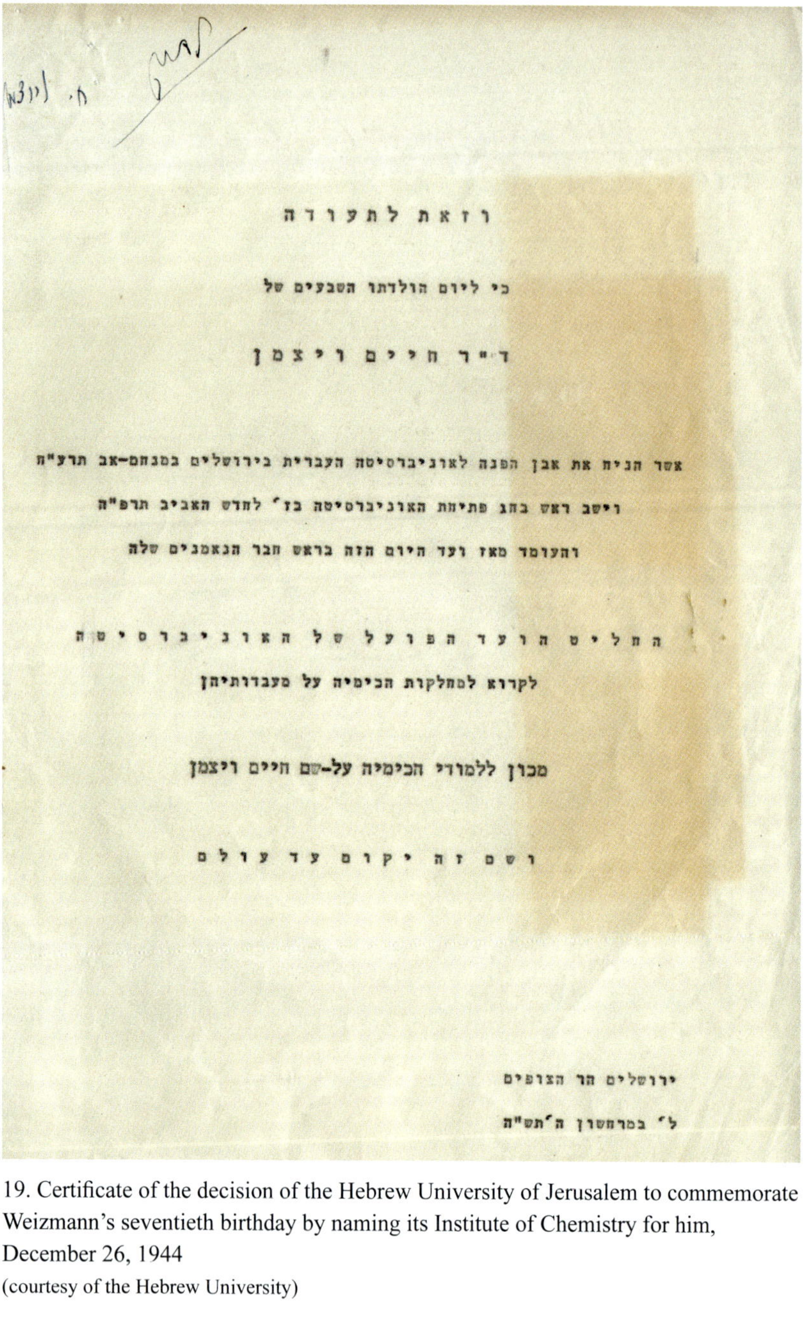

וזאת לתעודה

כי ליום הולדתו השבעים של

ד"ר חיים ויצמן

אשר הניח את אבן הפנה לאוניברסיטה העברית בירושלים במנחם-אב תרע"ח

וישב ראש בחג פתיחת האוניברסיטה בז' לחדש האביב תרפ"ה

והעומד מאז ועד היום הזה בראש חבר הנאמנים שלה

החליט הועד הפועל של האוניברסיטה

לקרוא למחלקות הכימיה על מעבדותיהן

מכון ללמודי הכימיה על-שם חיים ויצמן

ושם זה יקום עד עולם

ירושלים הר הצופים

ל' במרחשון ה'תש"ה

19. Certificate of the decision of the Hebrew University of Jerusalem to commemorate Weizmann's seventieth birthday by naming its Institute of Chemistry for him, December 26, 1944
(courtesy of the Hebrew University)

20. The Chaim Weizmann Institute of Chemistry building on Mt. Scopus
(courtesy of the Hebrew University)

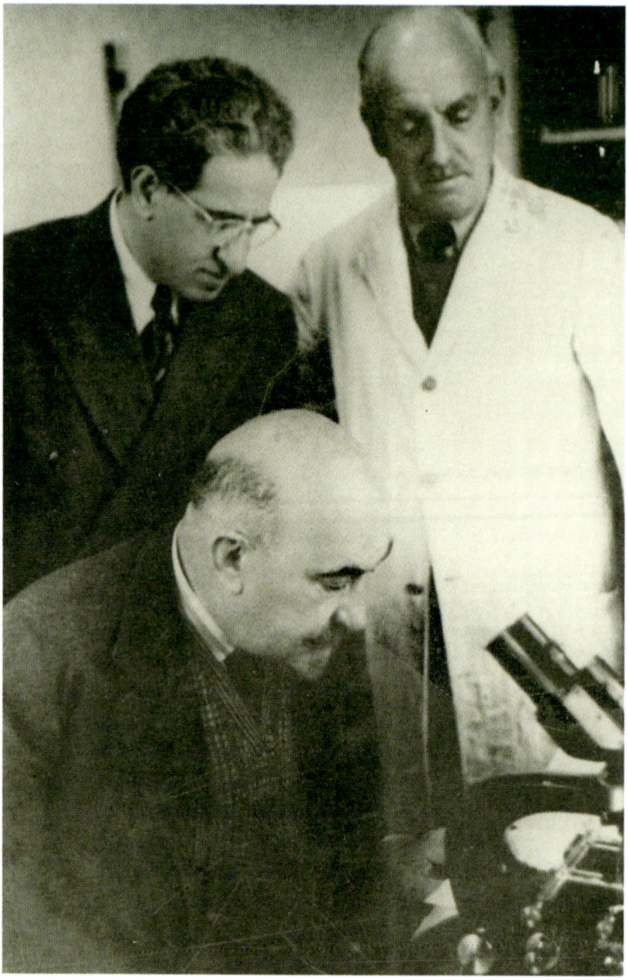

21. In Weizmann's laboratory at the Sieff Institute. From left to right: Meir Weisgal, Weizmann (seated), and Weizmann's longtime laboratory assistant, Harold Davies
(courtesy of the CWA)

22. Invitation to a fundraising dinner for the Weizmann Institute of Science, New York, January 27, 1945
(courtesy of the CWA)

23. Weizmann in his laboratory at the Sieff Institute, 1947
(courtesy of the CWA)

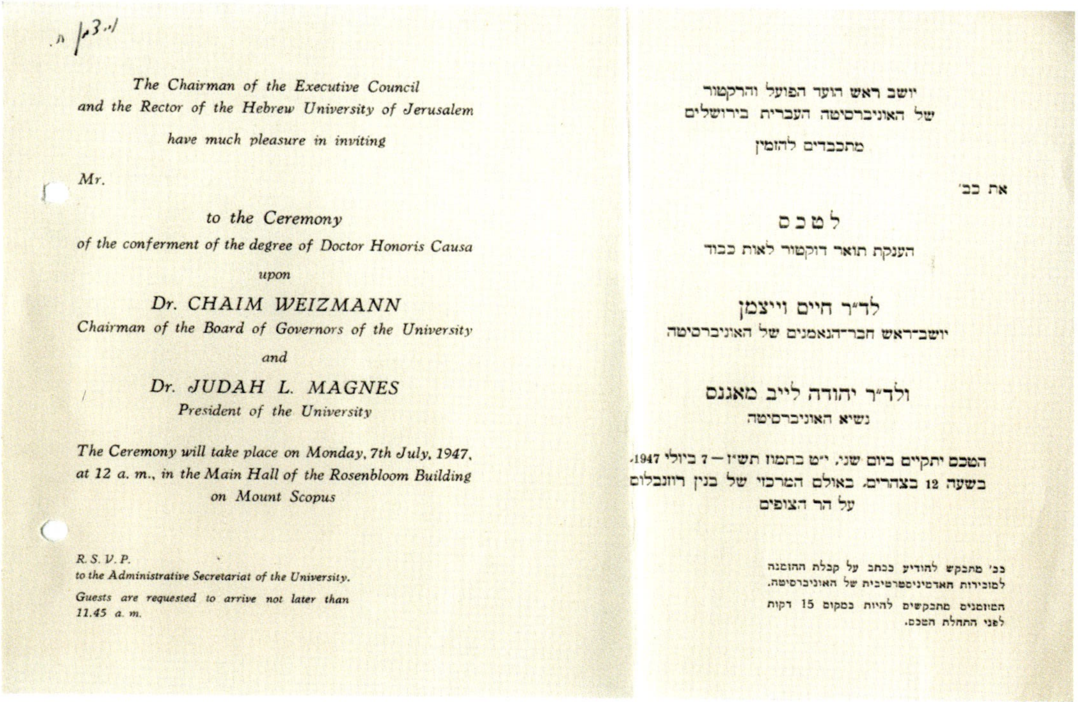

The Chairman of the Executive Council
and the Rector of the Hebrew University of Jerusalem
have much pleasure in inviting

Mr.

to the Ceremony
of the conferment of the degree of Doctor Honoris Causa
upon
Dr. CHAIM WEIZMANN
Chairman of the Board of Governors of the University
and
Dr. JUDAH L. MAGNES
President of the University

The Ceremony will take place on Monday, 7th July, 1947,
at 12 a. m., in the Main Hall of the Rosenbloom Building
on Mount Scopus

R. S. V. P.
to the Administrative Secretariat of the University.

Guests are requested to arrive not later than
11.45 a. m.

יושב ראש הועד הפועל והרקטור
של האוניברסיטה העברית בירושלים
מתכבדים להזמין

את כב׳

לטכס
הענקת תואר דוקטור לאות כבוד

לד״ר חיים וייצמן
יושב־ראש חבר־הנאמנים של האוניברסיטה

ולד״ר יהודה לייב מאגנס
נשיא האוניברסיטה

הטכס יתקיים ביום שני, י״ט בתמוז תש״ז — 7 ביולי 1947,
בשעה 12 בצהרים, באולם המרכזי של בנין רוזנבלום
על הר הצופים

כב׳ מתבקש להודיע בכתב על קבלת ההזמנה
למזכירות האדמיניסטרטיבית של האוניברסיטה.

המוזמנים מתבקשים להיות במקום 15 דקות
לפני התחלת הטכס.

24. Invitation to the ceremony for awarding honorary doctorates by the Hebrew University to Weizmann and Judah Magnes, to take place on July 7, 1947
(courtesy of the Hebrew University)

25. Weizmann and Magnes (right, seated) receiving their honorary doctorates from Sir Leon Simon, July 7, 1947
(courtesy of the Hebrew University)

26. Giulio (Yoel) Racah with Weizmann at the ceremony conferring his honorary doctorate, July 7, 1947
(courtesy of the Hebrew University)

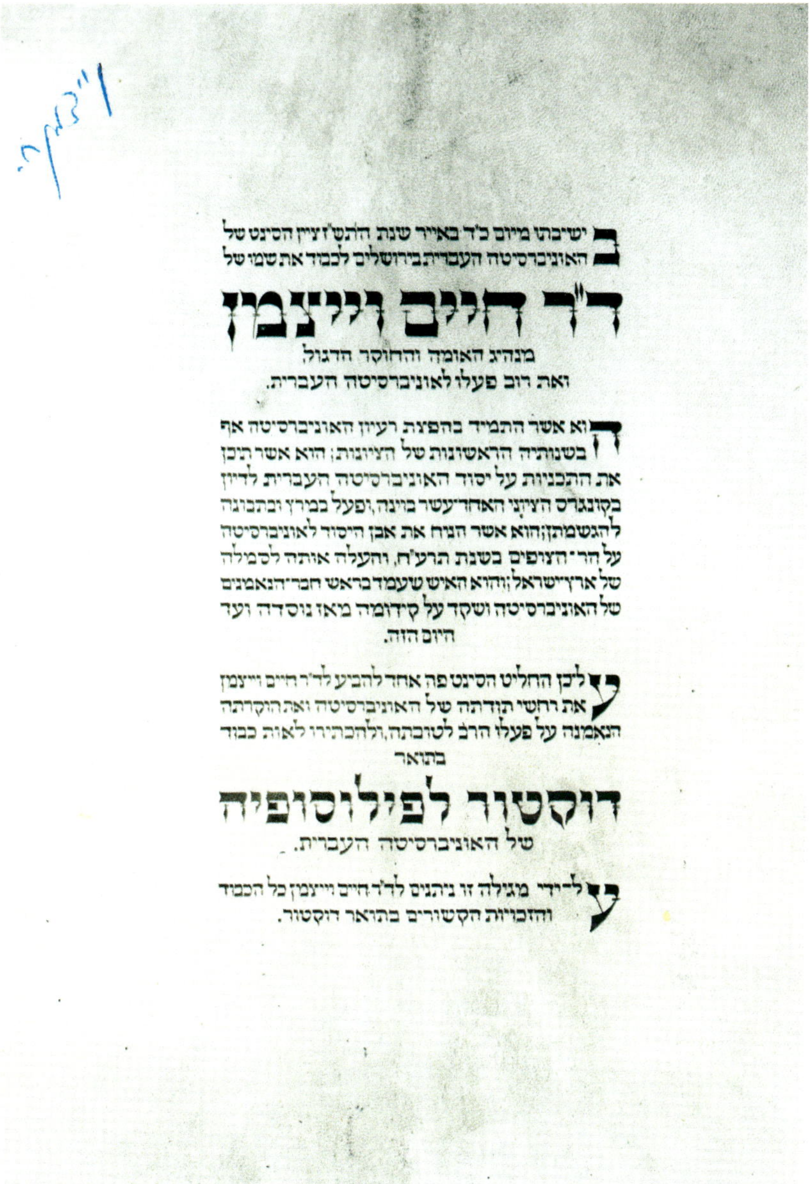

בישיבתו מיום כ"ד באייר שנת התש"ז ציין הסינט של
האוניברסיטה העברית בירושלים לכבד את שמו של

ד"ר חיים וייצמן

מנהיג האומה והחוקר הדגול,
ואת רוב פעלו לאוניברסיטה העברית.

הוא אשר התמיד בהפצת רעיון האוניברסיטה אף
בשנותיה הראשונות של הציונות; הוא אשר תיכן
את התכניות על יסוד האוניברסיטה העברית לדיון
בקונגרס הציוני האחד־עשר בוינה, ופעל במרץ ובתבונה
להגשמתן; הוא אשר הניח את אבן היסוד לאוניברסיטה
על הר־הצופים בשנת תרע"ח, והעלה אותה לסמלה
של ארץ־ישראל; והוא האיש שעמד בראש חבר־הנאמנים
של האוניברסיטה ושקד על קידומה מאז נוסדה ועד
היום הזה.

על כן החליט הסינט פה אחד להביע לד"ר חיים וייצמן
את רחשי תודתה של האוניברסיטה ואת הוקרתה
הנאמנה על פעלו הרב לטובתה, ולהכתירו לאות כבוד
בתואר

דוקטור לפילוסופיה

של האוניברסיטה העברית.

על־ידי מגילה זו ניתנים לד"ר חיים וייצמן כל הכבוד
והזכויות הקשורים בתואר דוקטור.

27. The honorary doctoral diploma awarded to Weizmann by the Hebrew University in 1947
(courtesy of the CWA)

28. Weizmann at the opening ceremony of the Weizmann Institute of Science, November 2, 1949 (courtesy of the CWA)

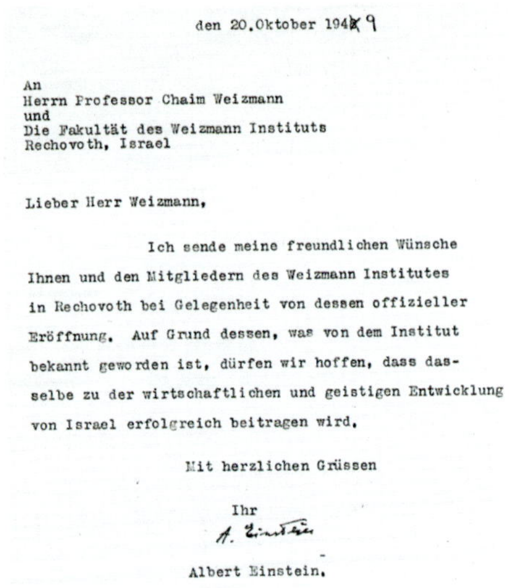

den 20.Oktober 194~~8~~ 9

An
Herrn Professor Chaim Weizmann
und
Die Fakultät des Weizmann Instituts
Rechovoth, Israel

Lieber Herr Weizmann,

Ich sende meine freundlichen Wünsche Ihnen und den Mitgliedern des Weizmann Institutes in Rechovoth bei Gelegenheit von dessen offizieller Eröffnung. Auf Grund dessen, was von dem Institut bekannt geworden ist, dürfen wir hoffen, dass dasselbe zu der wirtschaftlichen und geistigen Entwicklung von Israel erfolgreich beitragen wird.

Mit herzlichen Grüssen

Ihr

A. Einstein

Albert Einstein.

29. Letter from Albert Einstein congratulating Weizmann on the establishment of the Weizmann Institute of Science, October 20, 1949 (courtesy of the CWA)

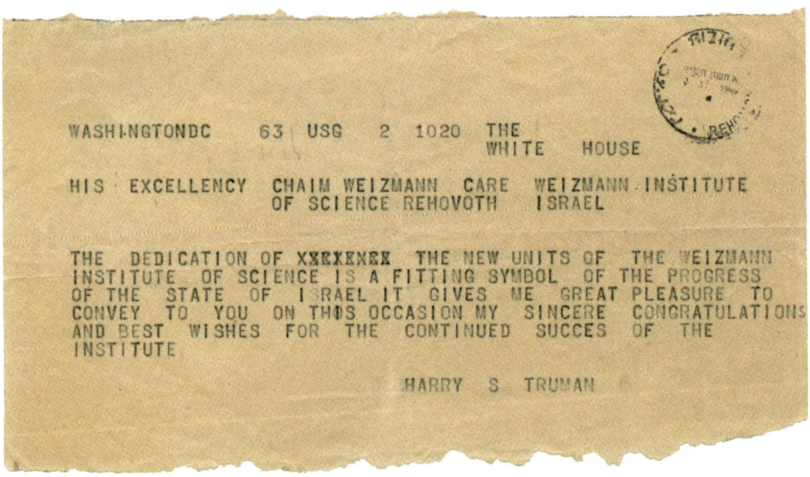

WASHINGTONDC 63 USG 2 1020 THE WHITE HOUSE

HIS EXCELLENCY CHAIM WEIZMANN CARE WEIZMANN INSTITUTE OF SCIENCE REHOVOTH ISRAEL

THE DEDICATION OF XXXXXXXX THE NEW UNITS OF THE WEIZMANN INSTITUTE OF SCIENCE IS A FITTING SYMBOL OF THE PROGRESS OF THE STATE OF ISRAEL IT GIVES ME GREAT PLEASURE TO CONVEY TO YOU ON THOS OCCASION MY SINCERE CONGRATULATIONS AND BEST WISHES FOR THE CONTINUED SUCCES OF THE INSTITUTE

HARRY S TRUMAN

30. Telegram from President Harry Truman congratulating Weizmann on the establishment of the Weizmann Institute of Science, November 2, 1949 (courtesy of the CWA)

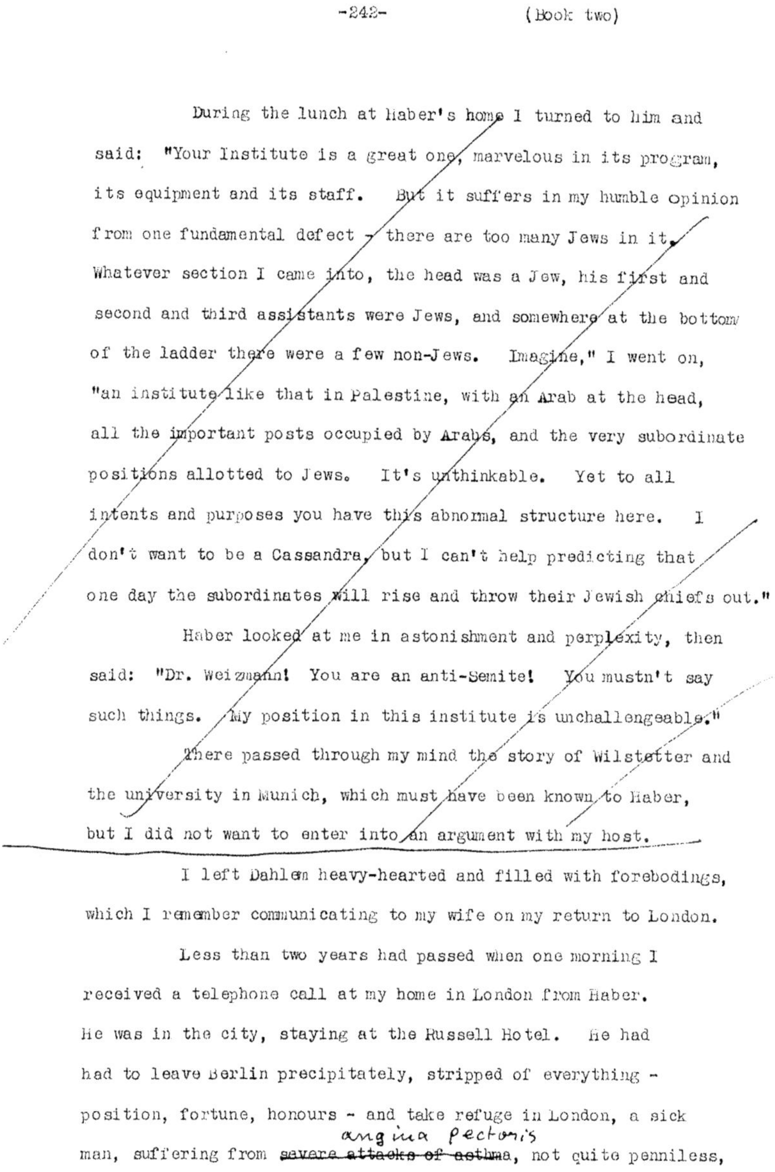

-242- (Book two)

During the lunch at Haber's home I turned to him and said: "Your Institute is a great one, marvelous in its program, its equipment and its staff. But it suffers in my humble opinion from one fundamental defect - there are too many Jews in it. Whatever section I came into, the head was a Jew, his first and second and third assistants were Jews, and somewhere at the bottom of the ladder there were a few non-Jews. Imagine," I went on, "an institute like that in Palestine, with an Arab at the head, all the important posts occupied by Arabs, and the very subordinate positions allotted to Jews. It's unthinkable. Yet to all intents and purposes you have this abnormal structure here. I don't want to be a Cassandra, but I can't help predicting that one day the subordinates will rise and throw their Jewish chiefs out."

Haber looked at me in astonishment and perplexity, then said: "Dr. Weizmann! You are an anti-Semite! You mustn't say such things. My position in this institute is unchallengeable."

There passed through my mind the story of Wilstetter and the university in Munich, which must have been known to Haber, but I did not want to enter into an argument with my host.

I left Dahlem heavy-hearted and filled with forebodings, which I remember communicating to my wife on my return to London.

Less than two years had passed when one morning I received a telephone call at my home in London from Haber. He was in the city, staying at the Russell Hotel. He had had to leave Berlin precipitately, stripped of everything - position, fortune, honours - and take refuge in London, a sick man, suffering from ~~severe attacks of asthma~~ angina pectoris, not quite penniless,

31. A page of the draft of Weizmann's autobiography with a passage struck out (see Section II in the chapter "Unpublished Passages")
(courtesy of the CWA)

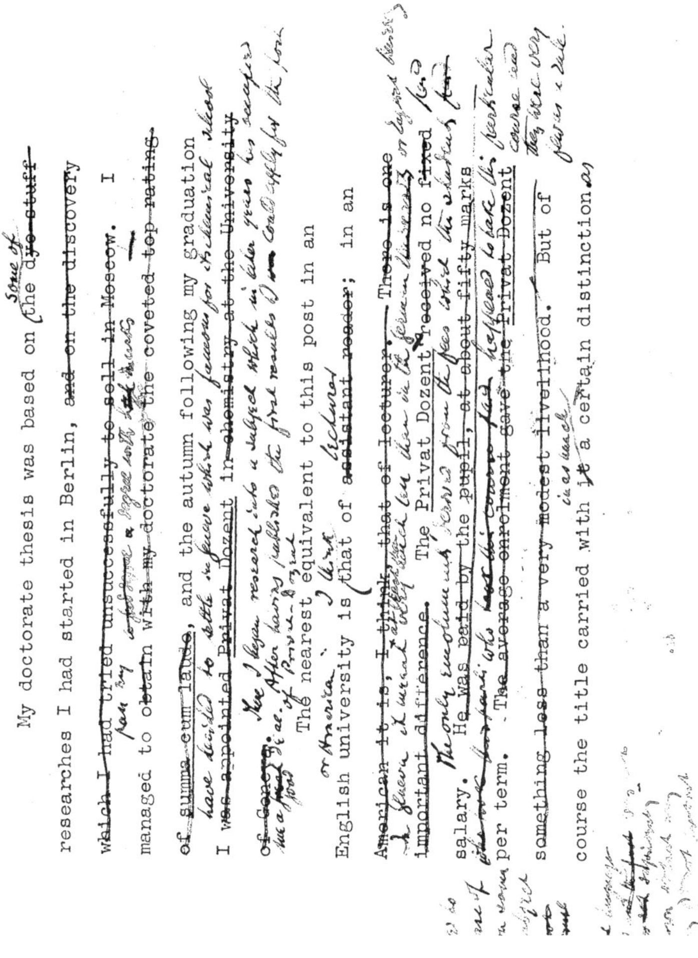

My doctorate thesis was based on the dye stuff
researches I had started in Berlin, and on the discovery
which I had tried unsuccessfully to sell in Moscow. I
managed to obtain with my doctorate the coveted top rating
of summa cum laude, and the autumn following my graduation
I was appointed Privat Dozent in chemistry at the University
of Geneva.
The nearest equivalent to this post in an
English university is that of assistant reader; in an
American it is, I think, that of lecturer. There is one
important difference. The Privat Dozent received no fixed
salary. He was paid by the pupil, at about fifty marks
per term. The average enrolment gave the Privat Dozent
something less than a very modest livelihood. But of
course the title carried with it a certain distinction.

32. Weizmann's corrections to an account of his scientific work in Switzerland; the revision was not incorporated into his autobiography. See Section V in the chapter "Unpublished Passages."
(courtesy of the CWA)

Patented Oct. 10, 1950 **2,525,249**

UNITED STATES PATENT OFFICE

2,525,249

MAKING ALKOXY ISOBUTYRIC ACIDS AND DERIVATIVES THEREOF

Charles Weizmann, Rehovoth, Israel, assignor to Polymerisable Products Limited, London, England, a British company

No Drawing. Application January 12, 1949, Serial No. 70,589

16 Claims. (Cl. 260—535)

1

The present application is a continuation-in-part of Ser. No. 591,856, filed May 3, 1945, now abandoned and a continuation-in-part of Ser. No. 640,129, filed Jan. 9, 1946, now Patent No. 2,490,109. This application may be considered as a consolidation of said two pending applications. This application contains certain matter from an earlier application 470,360 filed Dec. 28, 1942, now abandoned.

The present invention has reference to the making of certain alkoxy-isobutyric acids and their derivatives (e. g. esters).

I am aware that previously it was proposed to treat chloretone (a molecular compound or condensation product of chloroform and acetone) with a concentrated aqueous solution of caustic potash and to thereafter add methyl alcohol to the product so produced, and to then boil the mixture. Such a process was stated to yield methyl methacrylate, which could be separated from the other products present by fractional distillation. In said process, the chloretone and the aqueous caustic potash solution were indicated to be at room temperature, when mixed together. Such a process is inoperative to produce the result stated, since the said first reaction is so strongly exothermic that the reaction is violent and a too far-going decomposition of the product is produced.

In accordance with the present invention, the reaction is greatly slowed down by greatly limiting the amount of water present (if any), using a great bulk of an aliphatic alcohol in the reaction mixture, and previously strongly refrigerating the alcoholic solutions before mixing the same and regulating the temperature throughout the mixing operation. Thus the first reaction can be conducted without approaching the high temperatures at which such side reactions (decompositions) can take place.

The present invention has for its object, the safe production of alkoxy-isobutyric acids and their derivatives in yields approaching theoretical, and includes the process of making such bodies. Such substances are of importance for the production of methacrylic acid and its derivatives, in good yield.

The preparation of methacrylic acid derivatives by splitting off water or hydrohalogenic acid from α-hydroxy-isobutyric acid derivatives, or from α-halogeno-isobutyric acid derivatives, respectively, is a very violent reaction which therefore has usually been accompanied by partial polymerization of the product. The alkoxy-isobutyric acids and analogous compounds and

2

their derivatives (which have been made available by a novel and unexpected and unparallelled method) treated as described in the present specification, split off alcohol much more smoothly and give good yields of methacrylic acid derivatives and analogous acids, in a high state of purity, as set out below. The fundamental discovery is the fact that chloretone (readily produced by condensation of acetone and chloroform), or its homologues or derivatives, reacts with alcoholic sodium hydroxide or sodium alcoholate solutions or other corresponding alkali metal compounds, under proper heat control, to give alkali metal salts of alkoxy-isobutyric acids according to the following equation:

$$(CH_3)_2\underset{\displaystyle OH}{\underset{|}{C}}.CCl_3 + EtOH + 4NaOH =$$

$$(CH_3)_2\underset{\displaystyle OEt}{\underset{|}{C}}—COONa + 3NaCl + 3H_2O$$

"Et" means ethyl.

These acids give the usual derivatives, such as esters or phenylhydrazides. The esterification, for instance, can be carried out with alcohols in the presence of sulphuric acid. Under certain conditions, especially in the case of higher alcohols, this process is accompanied by a substantial conversion into methacrylic acid esters.

The alkoxy-isobutyric acid esters can be converted into the corresponding methacrylates by various methods, catalytic and non-catalytic ones, for instance by heating with oxalic acid or by passing the same over activated alumina. Another very convenient method consists in heating in the presence of phosphorus pentoxide which when carried out in the presence of the usual polymerization inhibitors, e. g. hydroquinone, pyrogallol or tannic acid, gives almost quantitative yields.

It is obviously not essential for the method that the alkyl groups in the alcohol and in the acid radical of the alkoxy-isobutyric acid esters are one and the same.

EXAMPLE 1

α-Methoxy-isobutyric acid

A cold solution of 448 grams (8 mols.) of potassium hydroxide in 250 c. c. (about 14 mols.) of water and 1000 c. c. (i. e. 796 grams or about 25 mols.) of methyl alcohol are placed in a vessel provided with a stirrer, a condenser and a superposed tank for the starting material. This is cooled to 0° C., or lower. Then one adds while stirring, a refrigerated solution (e. g. at 0° C.)

33. One of Weizmann's final patents, approved October 10, 1950 (courtesy of the CWA)

Joshua Jortner

Concluding Remarks

It is well known that Weizmann's scientific work formed the underpinning for his status as one of the modern world's leading scientist-statesmen and also contributed to his own self-awareness. He often spoke of his Zionist work and of the Zionist endeavor in terms borrowed from the world of science and the laboratory: "methodicalness"; "assessment of solutions"; "careful examination." His complementary approach to science and to Zionism rested upon the planning, adjustment, and realization of goals and of the means for their fulfillment; within it, science was endowed with a central role in building the land, renewing the Jewish people's national consciousness, and ensuring its normalization.

Contributors to This Volume

Ruth Arnon
Department of Immunology, Weizmann Institute of Science, and former President of The Israel Academy of Sciences and Humanities

Shlomo Avineri
Department of Political Science, The Hebrew University of Jerusalem, and The Israel Academy of Sciences and Humanities

Edward A. Bayer
Department of Biological Chemistry, Weizmann Institute of Science

Hedva Ben-Israel
Department of History, The Hebrew University of Jerusalem

Yigal Burstein
Department of Organic Chemistry, Weizmann Institute of Science

Joshua Jortner
School of Chemistry, Tel Aviv University, and former President of The Israel Academy of Sciences and Humanities

Shaul Katz
Institute of Contemporary Jewry, The Hebrew University of Jerusalem

Benjamin Z. Kedar
Department of History, The Hebrew University of Jerusalem, and former Vice President of The Israel Academy of Sciences and Humanities

Contributors to this Volume

Raphael Lamed
Department of Microbiology and Biotechnology, Tel Aviv University

Raphael Mechoulam
Faculty of Medicine, The Hebrew University of Jerusalem, and
The Israel Academy of Sciences and Humanities

Benny Morris
Department of Middle East Studies, Ben-Gurion University of the Negev

Issachar Unna
Racah Institute of Physics, The Hebrew University of Jerusalem

Shulamit Volkov
Department of History, Tel Aviv University, and The Israel Academy of Sciences and Humanities

Index of Names
Persons and Institutions

תוכן העניינים

תרגום לאנגלית: חיים ויצמן (Watzman)
עריכה והכנה לדפוס: דבורה גריינימן

התמונות באדיבות
יד חיים ויצמן, גנזך ויצמן, רחובות
ארכיון מכון ויצמן למדע
האוניברסיטה העברית בירושלים – הארכיון המרכזי

מסת"ב 978-965-208-225-1

©
כל הזכויות שמורות לאקדמיה הלאומית הישראלית למדעים, תשע"ז
סדר: אירית נחום
נדפס בישראל בדפוס איילון, ירושלים

חיים וייצמן

המדען, המדינאי ומדינאי־המדע

בעריכת

ב"ז קדר

ירושלים תשע"ז

האקדמיה הלאומית הישראלית למדעים

מנהיגות ומדע בישראל

כתבי האקדמיה הלאומית הישראלית למדעים

חיים וייצמן – המדען, המדינאי ומדינאי־המדע